Atmosphere-Ocean Modeling

Coupling and Couplers

Atmosphere-Ocean Modeling

Coupling and Couplers

Carlos R Mechoso

University of California Los Angeles, USA

Soon-Il An

Yonsei University, South Korea

Sophie Valcke

CERFACS, France

World Scientific

NEW JERSEY · LONDON · SINGAPORE · BEIJING · SHANGHAI · HONG KONG · TAIPEI · CHENNAI · TOKYO

Published by

World Scientific Publishing Co. Pte. Ltd.

5 Toh Tuck Link, Singapore 596224

USA office: 27 Warren Street, Suite 401-402, Hackensack, NJ 07601

UK office: 57 Shelton Street, Covent Garden, London WC2H 9HE

Library of Congress Cataloging-in-Publication Data

Names: Mechoso, Carlos R., 1942–   author. | An, Soon-Il, author. | Valcke, Sophie, author.
Title: Atmosphere-ocean modeling : coupling and couplers /
    Carlos R Mechoso, Soon-Il An, Sophie Valcke.
Description: New Jersey : World Scientific, [2021] | Includes bibliographical references and index.
Identifiers: LCCN 2021014318 (print) | LCCN 2021014319 (ebook) |
    ISBN 9789811232930 (hardback) | ISBN 9789811234460 (paperback) |
    ISBN 9789811232947 (ebook for institutions) | ISBN 9789811232954 (ebook for individuals)
Subjects: LCSH: Ocean-atmosphere interaction--Mathematical models. |
    Climatology--Mathematical models. | Multiscale modeling. | Computational complexity.
Classification: LCC GC190.5 .M44 2021 (print) | LCC GC190.5 (ebook) |
    DDC 551.5/246--dc23
LC record available at https://lccn.loc.gov/2021014318
LC ebook record available at https://lccn.loc.gov/2021014319

British Library Cataloguing-in-Publication Data
A catalogue record for this book is available from the British Library.

For any available supplementary material, please visit
https://www.worldscientific.com/worldscibooks/10.1142/12179#t=suppl

In praise of people leaders who encourage recognition of climate change and adoption of remedial actions. In hope that there were no others.

Preface

Although the equations that govern fluid motions were known since the beginning of the Industrial Revolution, numerical solutions of geophysical problems such as Earth's climate and its variability had to wait until the beginning of the Information Age. On the one hand, drawing a satisfactory picture of time changing, planetary scale phenomena involving both the atmosphere and the ocean as required for addressing the reliability of such solutions requires enormous amount of data. On the other hand, obtaining such solutions for periods of relevance to climate demands the power of supercomputers. In addition, it became clear around the middle of the last century that advances in technology did not suffice to guarantee that long-term integrations of the equations of fluid motion in the format available at the time would produce the stable solutions needed for climate applications. Hence, the equations themselves had to be modified, which required decades of work that was eventually successful. The stage was then set for the first step towards addressing the Earth's climate problem with numerical models: theory, data and technology came harmoniously together and numerical models of the atmosphere were produced with a recognized ability to simulate the evolution of the atmosphere if that of the underlying ocean was assumed to be known. Almost in parallel, numerical models of the ocean were produced and proved able to provide a convincing portrait of the oceanic circulation provided that the overlying atmosphere was assumed to be known.

The next daring step was to tackle the atmosphere-ocean interactions because these are essential contributors to the Earth's climate system. For example, let us say that we have developed a highly successful atmospheric model. To investigate why atmospheric behaviors were so different during two particular winters in the past, we could compare two model runs using already known ocean conditions. However, to predict a coming winter would require extrapolating ocean conditions into the future with a high risk of missing features that greatly influence atmospheric behaviors. Clearly the methodology to address this seasonal prediction problem was to obtain the oceanic conditions needed by the atmosphere from an ocean model running side by side and vice versa for the ocean. This is essentially the coupled atmosphere-ocean perspective. Initially, the behavior of coupled models showed that simply putting together models of the atmosphere and the ocean would not do. Some early attempts resulted in solutions so unrealistic that they were labeled "catastrophic". This was just a temporary setback and was quickly overcome. Finally, contemporary numerical models came about and they have a documented ability to produce realistic climates although with some resilient blemishes that remain to be conquered. Today, the increased interest for detail in climate model products by societal applications results in an ever increasing demand for faster computers as well as for further development of data retrieval, sorting, and manipulation systems. On the basis of the major scientific and technological challenges that have to be overcome, we believe that the numerical simulation of climate is one of the defining scientific achievements of the Information Age.

The emphasis of the present book is on the methodologies for coupling numerical models of the atmosphere and the ocean given the different levels of complexity of the models that are coupled, rather than on the models themselves. Coupling methodologies can even determine the failure or success of the model for an intended application. There is abundant literature describing numerical models of the atmosphere and oceans, their characteristics, performance, and successes as well as difficulties. Much less has been done in putting together a unifying view of the scientific and technical aspects of the coupling and participating couplers. In this regard, the present book intends to fill a void in the current literature by presenting a basic and yet rigorous treatment of how models of the atmosphere and the ocean are put together into a coupled system.

The presentation starts by reviewing the atmosphere-ocean processes that exchange information and by introducing the models that will be mentioned. These comprise sets of equations that can be solved analytically, hybrid arrangements in which one model component is a conceptual one while the other is a highly detailed representation, and others in which both ocean and atmosphere components consist of the most advanced algorithms to date. The chapters that follow examine the models and coupling assumptions. We have organized these chapters based on increased complexity of designs and breath of applications in climate studies, i.e. from methodologies addressing concepts to very detailed frameworks intended for detailed prediction of events that impact human activity. Such a strategy allows for a better recognition of the modeling issues as well as for the gradual introduction of outstanding climate events investigated with numerical models. The strategy also allows for a better appreciation of the many aspects to consider in putting component models together vis-à-vis the coupling problem. Two full chapters are dedicated to current efforts on the development of generalist couplers and coupling algorithms used by institutions for climate research and prediction all over the world.

We have followed the standard current thinking on modeling climate according to which different physical processes have come to be examined independently and even by different research groups while falling together in a superstructure that carries out interconnections and mutual dependences. In this framework, the notion of time sequencing is implied according to which the execution of one process occurs after another one is completed. It is obvious, however, that Nature does not work this way. To form clouds, for example, winds are not frozen in time. Winds at the ocean's surface change at the same time that ocean currents transport heat. In our view, future developments will blur the artificial division of climate models as a collection of different tasks. We are already seeing glimpses of this new paradigm as modern couplers may transfer full grid information instead of interface conditions. Time will tell whether this blurring of artificial boundaries is an aspirational aim ever to be approached but never to be reached. A principal goal of this book is to motivate graduate students to help accelerate the unifying undertaking. Thus, the presentation style conforms to a textbook for advanced undergraduate and graduate students. We hope that the text will be a useful reference for self-study by individual PhD students and researchers either coming into or specialized in the field, and appeal to a broader readership of individuals interested in a better understanding of Earth's environment as society prepares for climate change.

The authors would like to thank several colleagues for their careful reading of different chapters and for many useful comments: Jezabel Curbelo, Brian Medeiros and Chris Bretherton (Chapter 5); Stephen Griffies (Chapter 6); Eric Blayo, Anthony Craig, Rocky Dunlap, Moritz Hanke, Florian Lemarié, Robert Oehmke, and Li Liu (Chapters 8 and 9) Nicole Murray provided efficient technical help. S.-I. A. was supported by the National Research Foundation of Korea (NRF) grant funded by the Korean government (MSIT) (NRF-2018R1A5A1024958) and by the LG Yonam Foundation. S.V. gratefully acknowledges the European PRISM European project (EESD-1999-9) for the material produced on the PRISM revised ocean-atmosphere physical coupling interface described in Sec. 9.3.

Contents

Chapter 1

Atmosphere-Ocean Interactions and Feedbacks

1.1 Introduction

There is no doubt that solar insolation is the principal energy source for Earth's climate system. It is also known that climate does not respond passively to externally determined heating. The spatial and temporal distributions of various heat components in the atmosphere and oceans are determined by complicated interactions among several physical processes (such as radiation and convection) in ways that are strongly coupled with the motions of air and water. As a result, the atmosphere itself is cooled by the net radiation, being heated by heat fluxes at the surface. A major portion of this heat flux is in the form of latent heat, which is not recognized by the atmosphere as heating. The Earth's climate system is made substantially more complex by the presence of water substance. Oceans provide an almost infinite source of water, which is transferred to the atmosphere via surface evaporation. Oceans store and transport enormous amounts of heat.

1.2 Basic concepts

1.2.1 *The vertical temperature profile in the ocean*

Figure 1.1 contrasts idealized vertical profiles of ocean temperature in different locations and seasons. In broad terms one can recognize a layer near the surface where temperatures are approximately constant in the vertical; this is known as the oceanic mixed layer. Further down, the temperature decreases with depth in the layer known as the thermocline. In the deep ocean, temperatures again become almost constant in depth.

1.2.2 *The global distribution of sea surface temperature and precipitation*

The annual mean SST patterns in Fig. 1.2 show the expected higher values in the tropics decreasing away from the equator. Along the equatorial Pacific and

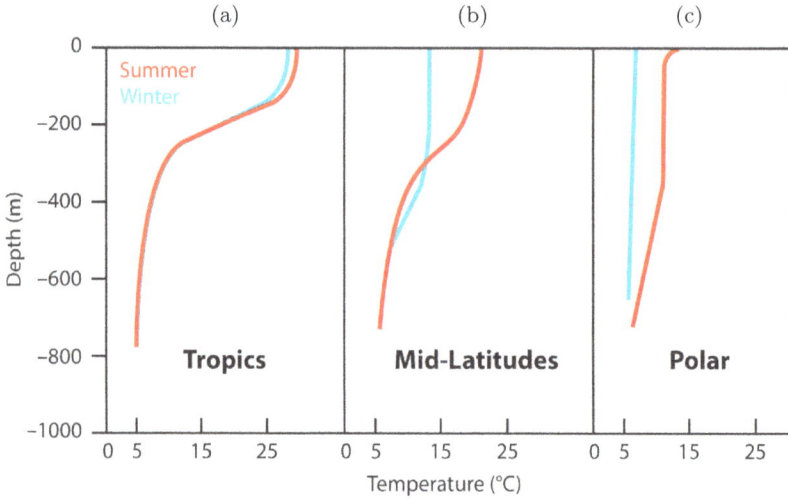

Figure 1.1.   Vertical profile of ocean temperature area averaged over a) Tropics, b) Mid-latitudes, and c) Polar regions. Red and blue curves indicate summer-mean and winter-mean conditions, respectively. (From: Exploring Our Fluid Earth https://manoa.hawaii.edu/exploringourfluidearth/ @University of Hawaii. Reprinted with permission from the Curriculum Research & Development Group.)

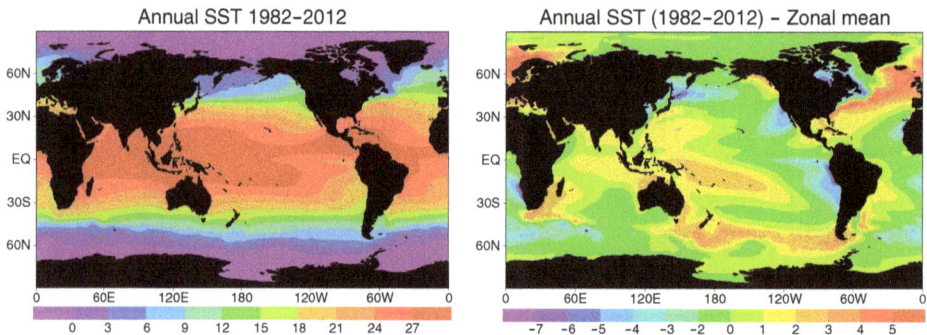

Figure 1.2.   Annual mean distribution of sea surface temperatures (SST) in the global ocean (left panel) and deviations from the zonal mean (right panel). Values have been averaged during the period 1982–2012. Units are °C. Data are from Reynolds [2002].

Atlantic, however, there are remarkable west-east asymmetries. Surface waters are warmer in the west than in the east. In the upper ocean, thermocline is deep in the west and shallow in the east as also shown in Fig. 1.3. The seasonal-mean pattern of SST has important interannual variability [Yu et al., 2021].

SSTs in both the eastern equatorial Pacific and Atlantic (left panel in Fig. 1.2) are highest in boreal spring when the solar irradiation is maximum in the tropics and the trade winds are weakest. Also, in this season, the thermocline

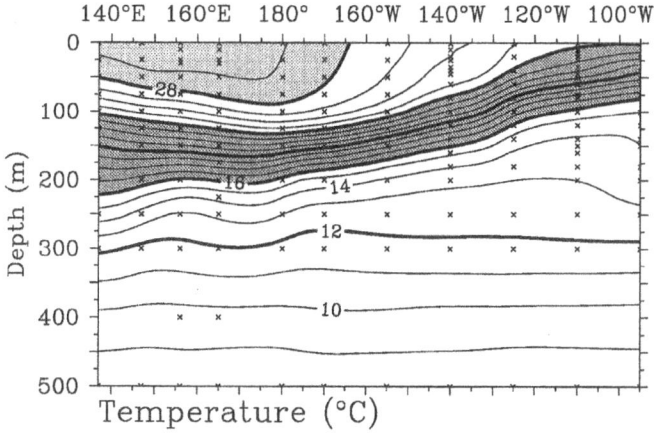

Figure 1.3. Longitude-depth plot of mean ocean temperature in the 2°S–2°N band of the Pacific. Units are °C. The 20°C isotherm depth is often taken as a proxy for thermocline depth. Data is from the Tropical Atmosphere Ocean (TAO) array for 1980–1996. (Figure 1 in McPhaden et al. [1998].)

is deeper in the east. Starting in April, the trade winds intensify, and a positive zonal pressure gradient develops along the equator, while SSTs in the eastern Pacific and Atlantic cool down and the thermocline shoals. The cooling in the Pacific is maximum in boreal fall, while that in the Atlantic is maximum in boreal summer. At these times, well-developed cold tongues appear in the eastern part of the oceans.

The deviations of SST about the zonal mean (right panel in Fig. 1.2) highlight the west-east asymmetries. The western parts of the Pacific and Atlantic are warmer than the eastern parts. This feature is less marked in the Atlantic due to the less extent of this ocean. Also, temperatures in the Southern Hemisphere tend to be lower than in the Northern Hemisphere. In the southern subtropics, SSTs are remarkably cold in the southeastern Pacific and Atlantic. These regions are characterized by strong coastal ocean upwelling and they are covered by extensive subtropical stratocumulus decks that shield the ocean surface from solar radiation. In subpolar latitudes, the North Atlantic where waters sink is warmer than the North Pacific where waters rise. Surface water in the sinking region is compensated by warm and salty water from lower latitudes.

The Pacific Ocean has experienced an overall warming trend since the 1880s. This feature has been primarily attributed to the increased atmospheric concentrations of greenhouse gases. A warming trend also has been detected in the Atlantic Ocean during the last century. This trend has been attributed to both natural and forced causes [Ting et al., 2009]. Although warming is apparent in both basins, it is geographically non-homogeneous.

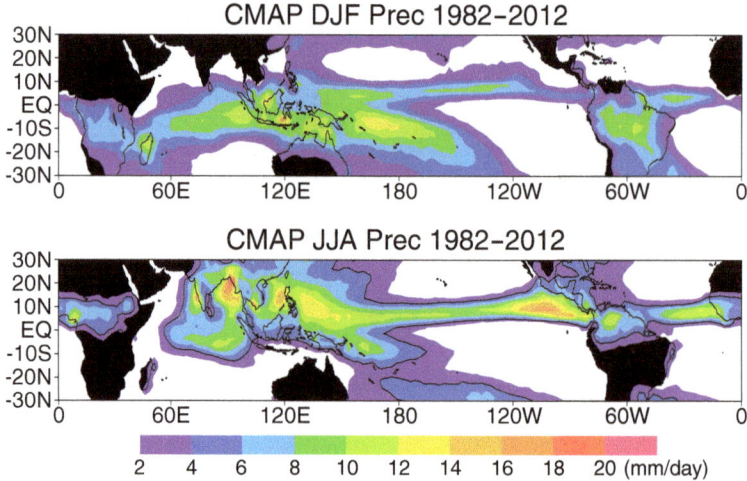

Figure 1.4. Annual mean distribution of precipitation over the global ocean for the boreal winter (December–February, upper panel) and the boreal summer (June–August, lower panel). Values have been averaged during the period 1982–2012. Units are mm/day. Data are from CPC Merged Analysis of Precipitation (CMAP) [Xie and Arkin, 1997].

Maps of tropical precipitation averaged for boreal winter (December–February; DJF) and summer (June–August; JJA) are shown in Fig. 1.4. An outstanding feature of precipitation in the tropical eastern Pacific and Atlantic is the Intertropical Convergence Zone (ITCZ). The ITCZ remains north of the equator during most of the year, i.e. does not follow the seasonal migration of insolation across the equator. In the Pacific, an analogue exists in the form of the South Pacific Convergence Zone (SPCZ), which is more strongly developed in the western Pacific and more prominent in boreal winter than in summer [Kiladis et al., 1989]. Ocean-atmosphere coupling processes are particularly critical for the variability of the cold tongue-ITCZ complex in the region, where the thermocline depth is shallow and ocean advection has a strong influence on SSTs.

The strong contrast between the panels in Fig. 1.4 reflects the development of continental monsoons during the warm seasons. The DJF map shows an arch of strong precipitation starting west on the South African monsoon, continuing east across the southern Indian Ocean to the strongest values in the southeast Asia-northern Australia region and ending with the SPCZ. The South American monsoon system is clearly seen also. The JJA panel captures the strongest monsoon systems on Earth over India and East Asia. Other monsoons are apparent over West Africa and Central America extending south towards equatorial South America. The Indian Ocean south of the equator also receives a large amount of precipitation in JJA.

GODAS Wind Stress, 1982-2004 Ann

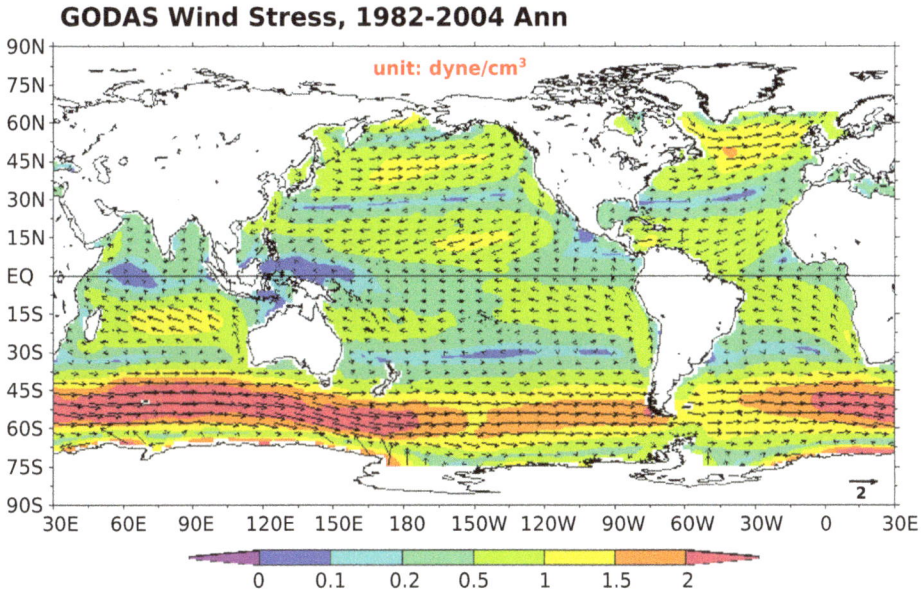

Figure 1.5. Annual-mean wind stress over the oceans for the periods of 1982–2004. Units are dyne cm^{-2}. Scale vectors are in the lower right corner. Data are from NCEP Global Ocean Data Assimilation System (GODAS) [Behringer et al., 1998; Behringer and Xue, 2004].

1.2.3 *Momentum and heat fluxes at the atmosphere-ocean interface*

In the mid latitudes, the surface flow over the oceans (Fig. 1.5) shows large scale circulations with westerlies on the poleward side and easterlies on the equatorward side. These easterlies are referred to as the trade wind systems and have an equatorward component in both hemispheres. Further poleward, the North Pacific and Atlantic are characterized by stationary low-pressure regions known as the Aleutian and Icelandic lows, respectively. In the Southern Ocean mid-latitude westerly winds drive the Antarctic Circumpolar Current (ACC), which flows eastward around the Antarctic continent.

The annual-mean net heat water flux (precipitation minus evaporation; Fig. 1.6) shows strong positive values in the regions of high precipitation highlighted in Fig. 1.4.

1.2.4 *Net energy fluxes at the atmosphere-ocean interface*

Figure 1.7 is a schematic of the energy fluxes in the atmosphere and the ocean and the exchange between them. The major topic of the present book is on the way in which climate models compute and transfer the fluxes at

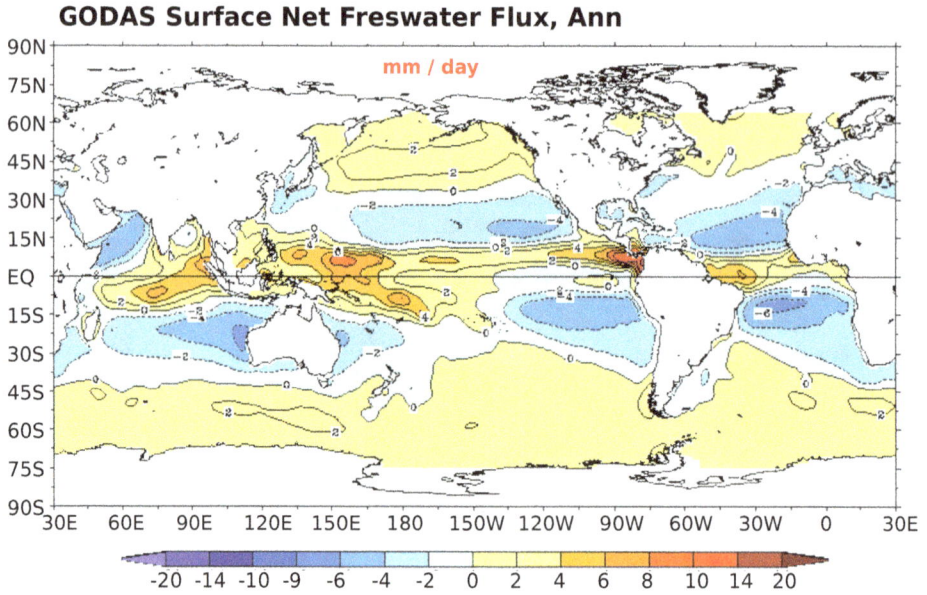

Figure 1.6. Annual-mean net freshwater water flux (precipitation minus evaporation) over the oceans. Units are mm/day. Data are from GODAS.

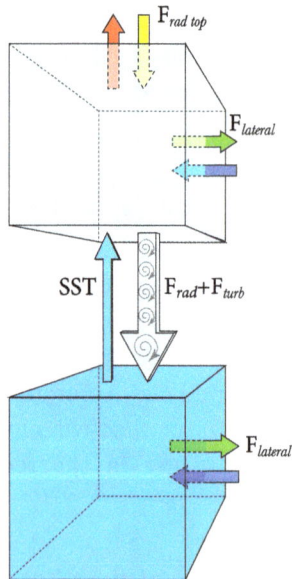

Figure 1.7. Schematic of the energy fluxes in the atmosphere and ocean, and of the exchanges at their interface. Symbols are explained in the text. The arrows emanate from the system component that plays the key role in the generation of the exchanged quantity.

the common interface of the two media. When atmosphere and ocean are represented by numerical models, these exchanges are referred to as "coupling" and the part of software that ensures the most precise and efficient possible transfers are referred to as "couplers". Figure 1.7 includes the fluxes associated with turbulence processes (momentum, sensible and latent heat). The other component in the schematic represents the exchanges of energy associated with radiative processes (F_{rad}). These radiative processes depend on emissions, absorptions and transmissions by the different gases that form the atmosphere as well as the ocean water, and on reflective properties of the surfaces. F_{rad} also depends on the microphysical and macrophysical properties of clouds, which are arguably the most challenging components to capture with numerical models. The schematic also includes exchanges of energy determined by internal processes within the atmosphere and oceans ($F_{lateral}$), such as advection and diffusion. Finally, SST is transferred from the ocean to the atmosphere.

Estimates of the net heat flux at the ocean surface are presented in Fig. 1.8, in which positive (negative) values represent ocean gain (loss). This complex field and the wind stress field are both the drivers and the consequence of the oceanic circulation. The ocean gains heat in the upwelling regions of the tropics and loses heat in the extratropics. Heat is lost in regions of the extratropics where the major oceanic current along the eastern coast of continents transport warmer water from the tropics. Heat fluxes in these regions are also affected by overflowing cold air from the continents. Another region of heat gain is seen in the North Pacific while the North Atlantic hints as another region of gain. The field shown in Fig. 1.8 results from contributions by different processes,

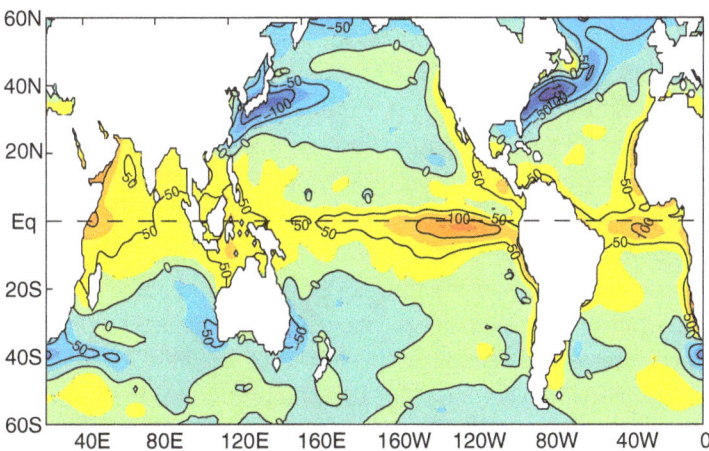

Figure 1.8. Annual mean net heat flux across the ocean surface. Units are Wm^{-2}. (Figure 1 in Fedorov et al., 2007, after Da Silva et al., 1994. ©American Meteorological Society. Used with permission.)

both external and internal, of the ocean. The complexity evidenced by the figures in this section serves as a reminder of the great challenge faced by coupled atmosphere-ocean models to be used in climate studies and projections.

1.3   Atmosphere-ocean feedbacks

Interactions between the atmosphere and ocean result in climate phenomena that neither the atmosphere nor the ocean on their own can produce. Much of these interactions are realized in feedback loops. Interactions in these loops can occur through processes and responses with different time scales. In this section we present some of the most important feedbacks of the climate system. As a reminder, we are interested in seasonal to interannual time scales such that important variabilities in very long time scales such as those involved in the melting of polar caps will not be addressed. We believe that such a strategy will not detract from our interest in coupling and couplers of models.

The surface heat fluxes may be driven by factors unrelated to SST (thereby acting as a forcing) or respond to SST anomalies (thereby acting as a feedback). Feedback processes are of great importance in the tropics. In the mid latitudes, the atmosphere is largely insensitive to local SST anomalies, which reduces the possibilities for strong air-sea coupling [see, e.g., Kushnir and Held, 1996].

1.3.1   *Bjerknes feedback*

The tropical Pacific SST is characterized by a remarkable cold tongue in the eastern equatorial region and a warm pool over the western basin and Maritime Continent, giving rise to a strong gradient along the equator (Figs. 1.2 and 1.3). Bjerknes argued that the east-to-west SST gradient that characterizes the equatorial Pacific can drive a direct thermal circulation of the atmosphere with ascending motion to the west and descending motion to the east (the Walker circulation) [Bjerknes 1966, 1969]. A small perturbation in either the Walker circulation or zonal SST gradient may become unstable and grow further, giving rise to the seesaw-type variations on interannual timescales in sea level pressure [the Southern Oscillation; Walker, 1923] and in SST ("El Niño", see Ch. 3). The instability is established through a positive feedback between variations in the equatorial trade winds and underling SSTs over the Pacific. A similar, albeit weaker, chain of processes seems to develop in the equatorial Atlantic giving rise to the Atlantic Niño. This feedback is referred to as the "Bjerknes feedback".

There is observational evidence for the different components of the Bjerknes feedback. Deppenmeier et al. [2016] focused on the tropical Atlantic and subdivided the feedback into three components by inspecting the relationships among, (1) SST anomalies in the eastern equatorial basin and zonal wind stress anomalies in the western basin, (2) wind stress anomalies in the western tropical basin and ocean heat content anomalies in the eastern equatorial basin, and (3) heat content anomalies in the eastern tropical basin and overlying SST anomalies. The patterns of these relationships in the ERA-Interim [Dee et al., 2011] and ORAS4 [Balmaseda et al., 2013] reanalysis were compared with those simulated by an ensemble of CMIP5 models in the historical scenario. The results show that components (1) and (2) of the feedbacks are simulated relatively well, while component (3) is weaker in the models and its pattern is more spread out about the equator. This latter feature was attributed to the erroneous vertical oceanic stratification in the eastern basin, which points to model difficulties with ocean mixing and/or the atmospheric forcing.

1.3.2 *Wind evaporation SST feedback*

Another important feedback in the tropics involves the Hadley circulation and the meridional SST gradient. The northward SST gradient associated with a perturbation at the equator will induce a northward pressure gradient that in turn will drive a southerly cross-equatorial wind anomaly. As the anomalous wind continues to flow northward away from the equator, the Coriolis effect acts to veer the wind toward the northeast, opposite to the direction of the background northeasterly trade wind. This results in a decrease in the latent heat release from the ocean to the atmosphere, generating a surface warming north of the equator and amplifying the initial warming. This feedback is referred to as the "Wind–Evaporation–SST feedback" [WES; Xie, 1996; Chang et al., 1997]. Vimont et al. [2001, 2003] proposed a seasonal footprinting mechanism (SFM), which explained how the wintertime midlatitude atmospheric variability influences the following spring and summer tropical atmosphere via the WES feedback. Mahajan et al. [2011], based on experiments with a numerical climate model, suggested that the WES feedback is not essential in the propagation of an artificial cooling signal from high-latitudes to the deep tropics. They also argued, in contrast, that the WES feedback is responsible for amplifying SST and atmospheric anomalies in the latitude band between about 10°S and 10°N (deep tropics).

1.3.3 *Ekman feedback*

In the deep tropics, a cross-equatorial wind anomaly driven by a cross-equatorial SST gradient is reinforced by a positive Ekman upwelling anomaly in the colder hemisphere, constituting a positive feedback. This feedback is referred to as the "Ekman feedback", which plays a role in the maintenance of the asymmetry of the ITCZ/cold tongue complex in the eastern equatorial Pacific and Atlantic [Chang and Philander, 1994].

1.3.4 *Water vapor feedback*

Water vapor in the air absorbs infrared radiation but is relatively transparent to solar radiation (i.e., it is a greenhouse gas). When water vapor is present in the air, the effective emission height in the atmosphere is higher than otherwise. For a given temperature profile, a higher emission height implies a lower emission temperature and hence less emitted radiation. In order to balance incoming radiation, therefore, the temperature at the effective emission height must increase. The constancy of the lapse rate requires a higher surface temperature [Held and Soden, 2000]. Warmer air can hold more water vapor, which implies a further increase in emission altitude, and so on. This process defines the powerful water vapor positive feedback. It is accepted that this feedback may amplify the response to a perturbation in water vapor by as much as a factor of two, although there is some debate on the magnitude of amplification.

A stratospheric water vapor feedback has also been proposed [Dessler et al., 2013]. Increasing water vapor entering the stratosphere through tropopause as temperatures rise in the troposphere leads to warming. The estimated warming through stratospheric water vapor feedback is about $0.3\,\mathrm{W/m^2/K}$.

1.3.5 *Cloud-SST feedback*

Clouds affect the net surface radiative flux at the ocean surface, exerting a direct impact on the surface energy budget and thus SST. The associated changes in cloud macrophysical and microphysical properties can either positively or negatively feedback onto the SST perturbation that produced these changes, leading to either amplified or dampened SST response. The effects of clouds on radiation is expressed by the Cloud Radiative Effect (CRE). CRE is the difference in upwelling upward radiation at the top of the atmosphere if there were no clouds and that when clouds are present [Ramanathan et al., 1989].

A well-known example of the positive cloud-SST feedback is the feedback between low-level marine stratus clouds and SST over the eastern Pacific and Atlantic coastal upwelling regions. Warmer SSTs imply less clouds and more solar radiation reaching the ocean surface [Klein and Hartmann, 1993; Philander et al., 1996; Ma et al., 1996; Nigam, 1997; Huang and Hu, 2007].

1.3.6 *Thermocline feedback*

One of the most important components in Bjerknes feedback is thermocline feedback. When the mean thermocline is shallow, the perturbations by oceanic wave motion are more effective in influencing SST. For example, a slightly relaxed SST contrast between the tropical eastern and western Pacific drives westerly anomalous surface wind over the tropical central Pacific [Lindzen and Nigam 1987], which generates downwelling Kelvin waves. The eastward propagating downwelling Kelvin waves deepen the thermocline depth over the eastern Pacific. Because of this deepening, and especially over the eastern Pacific, anomalous upwelling water from the subsurface to the surface becomes warmer than normal. As a result, the eastern Pacific SST anomaly increases, and consequently it further reduces the SST contrast between eastern and western Pacific.

1.3.7 Ice-albedo feedback

Sea-ice releases latent heat of fusion when it forms and absorbs latent heat of fusion when it melts. Ice has a very different reflectivity to sunlight (albedo) than open ocean or land areas. Hence, the larger the ice caps, the more sunlight will be reflected back to space. If the solar brightness is lowered slightly, the planetary temperature lowers slightly leading to an expansion of the ice caps. This causes the planet to be more reflective to sunlight, leading in turn to an enhanced cooling. The ice-albedo feedback was among the first examined in climate studies [e.g., Budyko, 1969; Sellers, 1969].

1.3.8 Land surface conditions feedback

Although our primary interest in this book is on the atmosphere-ocean interface we take a short detour and mention the atmosphere-land surface processes. The importance of these processes has been mostly supported by studies on climate sensitivity to albedo [e.g., Charney et al., 1979; Dirmeyer and Shukla 1994], soil moisture [e.g., Shukla and Mintz, 1982; Douville et al. 2001; Koster et al., 2009], other individual land variables such as surface roughness [Sud et al., 1988] and leaf area index (LAI) [e.g., Chase et al., 1996; Kang et al., 2007], and some combinations of these variables [e.g., Yasunari et al., 2006].

Evapotranspiration supplies moisture to the atmosphere over wet or vegetated land becoming a major component of the global water cycle. For global and regional studies with numerical models, the effects of soil moisture are represented by specialized algorithms such as the Simplified Simple Biosphere Model [SSiB; Xue et al., 1991, 1996]. SSiB takes into consideration surface turbulent fluxes of water and sensible heat as well as radiative transfer in the canopy and simulates diurnal and seasonal variations of albedo, canopy transpiration, and water interception loss. Vegetation morphology and canopy resistance and their seasonal changes are also taken into account in the computation. Sensitivity studies [e.g. Xue et al., 2010] have shown that simulations of global and regional precipitation with different representations of land surface processes could have very substantial differences resulting in large uncertainties in precipitation especially in monsoon regions.

1.3.9 *Reemergence*

Namias and Born [1970, 1974] suggested that the shallow oceanic mixed layers that form in late spring and early summer due to increased surface heating and weaker surface winds isolate the mixed layers of colder SSTs that form below in the previous winter. Stronger winds and decreased surface heating in the following winter, therefore, would entrain colder waters from below [Alexander and Deser 1995; Alexander et al., 1999; Zhao and Li, 2010]. This argument applies to the SST anomalies in such a way that, due to such a mechanism, an anomaly during a particular winter may have effects one year later [Geiss et al., 2020]. Consequently, this oceanic thermal inertia acts to integrate atmospheric forcing, and finally contributes to long-term climate variations such as the Pacific Decadal Oscillation [Newman et al., 2016].

Chapter 2

A Classification of Coupled Atmosphere-Ocean Models

2.1 Introduction

A vast assortment of numerical (mathematical) models is used in climate studies. In the present book we focus on models of the atmosphere-ocean system that are based on either simplified or comprehensive or equations of fluid motion, conservation of water and other substances, the laws of thermodynamics, and the physics of radiation transfer. This excludes the so-called energy balance models of the type developed by M. I. Budyko and independently by William Sellers in the late 1960s and early 1970s [Budyko, 1969; Sellers, 1969; North 1975]. We also exclude models that are one-dimensional in space, even though we recognize that such a framework has the potential to be useful in addressing fundamental modeling issues.

This chapter introduces the models we have selected for study. In presenting the models, we have opted for an organization based on increased complexity of designs and breadth of applications in climate studies, i.e. from methodologies addressing concepts to very detailed frameworks intended for detailed prediction of events that impact human activity (see Fig. 2.1). Such a strategy allows for a better understanding of the modeling issues as well as for the gradual introduction of outstanding climate events investigated with numerical models. The strategy also allows for a better appreciation of the many aspects to consider in putting component models together vis-à-vis the coupling problem.

We will start with "simpler climate models", which are broadly understood as those that include a minimum set of the major physical processes at work in a phenomenon selected for study. There can be as many simple models as basic questions to be resolved on climate and its variability. In the category of simpler models, we will consider separately the "conceptual" and "intermediate complexity models". Next, we will move to comprehensive models; these are referred to as general circulation models of the atmosphere and oceans (AGCMs and OGCMs, respectively). At this point we will examine "hybrid coupled models", which consist of an AGCM coupled to a simpler ocean model

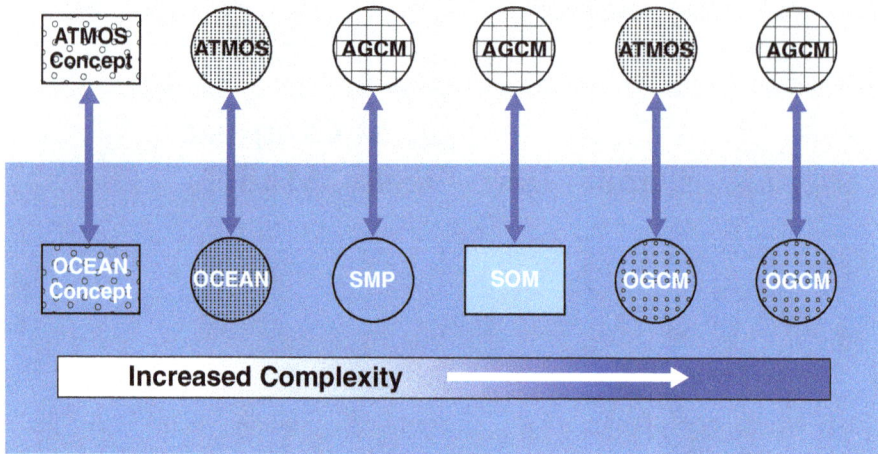

Figure 2.1. Schematic of coupled atmosphere-ocean models organized in terms of increased complexity from left to right. "Concept" indicates a conceptual model; "ATMOS" and "OCEAN" refers to simple models; "SMP" and "SOM" are acronyms for "Swamp Ocean Model" and "Slab Ocean Model", respectively; "AGCM" and "OCCM" are acronyms for atmosphere and ocean general circulation models, respectively.

(Type 1 or HCM1) or an OGCM coupled to a simpler atmospheric model (Type 2 or HCM2). Finally, we will turn to AGCMs coupled to OGCMs or fully coupled GCMs (CGCMs). Two full chapters are dedicated to coupling technologies and their use by modeling groups around the world.

2.2 Conceptual models of coupled atmosphere-ocean processes

In conceptual models, physical processes are represented on the basis of a bulk conceptualization of their effects. We have selected as examples of such models the most popular ones used in studies on El Niño/Southern Oscillation (ENSO; Suarez and Schopf [1988], Battisti and Hirst [1989], Jin [1997]). The validity of such models is generally assessed by their success in producing oscillatory behaviors that characterize ENSO, albeit the numerical value of the oscillation period may depend on assumed parameters having values that differ from those suggested by observations. *Conceptual models incorporate special idealized expressions for the coupling between atmosphere and oceans. Such expressions depend on ad-hoc parameters, and the methodology for research with such models addresses their sensitivity to these parameters.*

Descriptions of the conceptual models which we have selected and examples of their applications are given in Ch. 3.

2.3   Models of intermediate complexity and ENSO prediction

This category includes models that share two important characteristics: (1) the atmosphere and ocean components are both based on drastically simplified versions of dynamical and physical processes, and (2) they are able to make ensemble predictions for months in advance with relatively modest computational resources [Cane et al., 1986; Zebiak and Cane, 1987]. The application focus is on the skillful prediction of ENSO as well as the better understanding of fundamental processes in ENSO evolution. *Intermediate complexity models also incorporate specialized expressions for coupling between the atmosphere and oceans. However, the formulation of these expressions may explicitly consider observational data.*

Descriptions of the intermediate complexity models which we have selected and examples of their applications to ENSO prediction are given in Ch. 4.

2.4   AGCMs coupled to simpler ocean models (HCM1s)

AGCMs synthesize a number of mutually interacting processes that are deemed as essential for the general circulation of the atmosphere with its many interactions and feedbacks. Numerical methods are applied to explicitly solving the equations governing fluid motion using parameterizations of processes not resolved by the models' grids. The air is represented as flowing above realistic representations of mountains as well as vegetated areas and deserts while forming jet streams and other wind systems. A hydrological cycle of evaporation-transport-precipitation of water vapor is simulated, and clouds of different types affecting energy transfers. AGCMs have complex codes that require computer resources at the high end of those available at any time. For climate applications AGCMs are coupled to models of the ocean. We will refer to an AGCM coupled to a simple ocean model as a hybrid coupled model of Type 1 (HCM1s). *In an HCM1, forcing of the ocean is provided by complex algorithms that are integral parts of the AGCM, while the oceans return the sea surface temperature (SST) according to simplified ocean dynamics and thermodynamics. In this way, coupling uncertainties are greatly reduced formally.*

Descriptions of AGCMs are given in Ch. 5. This is followed by examples of HCM1 configurations and applications.

2.4.1   *AGCMs coupled to a swamp ocean model*

A swamp ocean means a wet surface with zero heat capacity. Such a model uses energy balance formulations to determine the SST in an ocean with no currents. Geographically, the wet surface can be assumed for the whole extent of the World Ocean or for particular ocean basins, and even more ideally, the entire

Earth's surface in a configuration that is referred to as an Aquaplanet [Medeiros et al., 2016]. A swamp ocean model, therefore, consists of a surface that is always saturated and able to provide the atmosphere with the required moisture whatever the demand is. The terminology originated in early studies by S. Manabe and collaborators [e.g., Manabe and Wetherald, 1975]. One advantage of this type of ocean model is the instantaneous convergence to quasi-equilibrium for a given climate forcing because there is no heat storage effect and no slow dynamic adjustments.

2.4.2   *AGCMs coupled to a slab ocean model*

The slab or simplified ocean model (SOM) essentially represents the ocean mixed layer at every ocean point. The layer depth is specified as being either geographically variable or globally homogeneous, with values based on observational data. The layer temperature is the SST, which is calculated using the energy budget of the mixed layer with atmospheric fluxes provided by the AGCM and oceanic fluxes provided by an algorithm based on the observation [e.g., Manabe and Stouffer, 1980; Washington and Meehl, 1984; Wetherald and Manabe, 1988]. The SOM is dynamically passive and does not provide oceanic heat transports, therefore these are parameterized as climatological "Q fluxes" derived from a long control run of the same OGCM with forcing corresponding to an observed climatology. An additional relaxation term to an observed climatology is generally incorporated to the SST equation. The success of SOMs in simulating SST distributions and their seasonality with this type of parameterizations has been reported. However, the passive nature of SOMs limits their utility in studies of climate change sensitivity, in which oceanic energy transport is *per force* assumed to be invariant. Nevertheless, SOMs have proved useful in studies of weak air-sea coupled systems such as Madden–Julian Oscillation and typhoon/hurricanes, in which the atmosphere mainly interacts with a mixed-layer ocean.

2.5   OGCMs coupled to simpler atmospheric models (HCM2s)

Oceanic general circulation models (OGCMs) synthesize a number of mutually interacting processes that are deemed as essential for the general circulation of the world oceans with its many interactions and feedbacks. As in AGCMs, numerical methods are applied to explicitly solving the equations governing fluid motion using parameterizations of processes not resolved by the models' grids. In full OGCMs the flow of water is confined by continents with realistic coastlines. The water forms ocean currents and either sinks or upwells in special geographical regions. Sea-ice forms at the surface of polar oceans. OGCMs have complex codes and require computer resources that are as important and can be even more important than AGCMs. For climate applications OGCMs are coupled to models of the atmosphere. We will refer to an OGCM coupled to a simpler atmospheric model as a hybrid coupled model of Type 2 (HCM2s). *In an HCM2, the SST produced by the OGCM provides the atmospheric forcing of the ocean using algorithms based on statistical properties of the observed atmosphere. Use of these special models requires a very careful selection of research issues to be investigated both in terms of time and space scales.*

Descriptions of OGCMs and examples of HCM2s are given in Ch. 6. The simpler atmospheric models have been introduced in Chs. 3 and 4.

2.6 Coupled atmosphere-ocean GCMs (CGCMs)

Coupled atmosphere-ocean GCMs (CGCMs) are comprehensive numerical models used for research and prediction of climate variability (e.g., ENSO and anthropogenic climate change). For climate studies, coupled AGCMs and OGCMs incorporate representations of land, sea-ice and iceberg dynamics. CGCMs have two primary applications: (1) weather and climate prediction, and (2) investigations aimed to increase understanding of the climate system and its variability. An important CGCM application has been the simulation and prediction of climate change expected as a result of the changing atmospheric composition. The impact of increased greenhouse gases has been assessed by comparing model climates with different prescribed concentrations of those gases corresponding to agreed emission scenarios. In addition, CGCMs provide an experimental tool for studying scale and regional interactions and various positive and negative feedbacks. *CGCMs do not need additional algorithms to calculate coupling fields, although a relaxation term to an observed climatology may be added to some interface variables to minimize the model's drift. CGCMs provide sophisticated laboratories for testing how well we understand and can predict atmospheric and oceanic processes, including the impacts of varying surface conditions and changes in the atmospheric composition.*

Descriptions of CGCMs and examples of applications are given in Ch. 7.

2.7 Perspectives

As indicated in the previous section, climate predictions with CGCMs require to prescribe future concentrations of greenhouse gases. In order to eliminate this requirement, interactive biogeochemistry, including the carbon cycle, has been incorporated into CGCMs. The resulting tools are referred to as Earth System Models (ESMs). The development, testing, and optimization of ESMs in seen as a multidisciplinary task involving groups of researchers in different institutions. The science and computational challenges are daunting: Adding new components to an ESM must be done in an internally consistent context while the extraordinary computational demands of the codes require a high level of efficiency. Some issues in the coupling of ESM components are mentioned in Ch. 8 and 9.

Chapter 3

Conceptual Models of Interannual Variability

3.1 Introduction

This chapter is dedicated to coupled atmosphere-ocean models specially designed to examine the fundamental processes that determine the interannual variability in the tropical oceans. The prime example of application of such conceptual models is a better understanding of the dynamics of the El Niño-Southern Oscillation (ENSO) cycle. The models to be reviewed are the Recharge Oscillator [Jin, 1997] and the Delayed Oscillator [Schopf and Suarez, 1988; Battisti and Hirst, 1989], which will be referred as JIN97 and SSBH, respectively. These models have provided qualitatively plausible starting points for understanding ENSO, i.e. they give the simplest representation of the physics of the mixed sea surface temperature/ocean-dynamic mode with subsurface ocean temperature anomalies carrying the memory during the transition phases when little anomaly is visible at the surface. The presentation that follows highlights the hypotheses made to represent atmosphere-ocean interactions. A short discussion on the implied absence of physics due to the simplifications is presented. For example, both models miss the effects on the oscillation of ocean-mixed layer processes away from the equatorial eastern Pacific.

3.2 Basic aspects of ENSO

ENSO is the paradigm for atmosphere-ocean coupled variability. Starting with the atmosphere, we plot regressions of key quantities onto the Niño-3.4 SST index, which is defined as the sea surface temperature (SST) anomalies averaged over (5° N–5°S, 120°–170°W). Figure 3.1 shows the results for the winter season [December–February (DJF)] in the NCEP–NCAR reanalysis [see also Wallace et al., 1998].

The SST anomalies show positive values in the central and eastern equatorial Pacific (Fig. 3.1(a)). Precipitation anomalies are positive around the central equatorial Pacific with negative anomalies around them and largest values slightly to the west of the maximum SST anomalies (Fig. 3.1(b)). The

NCEP NINO3.4 Regression DJF 1982-2008

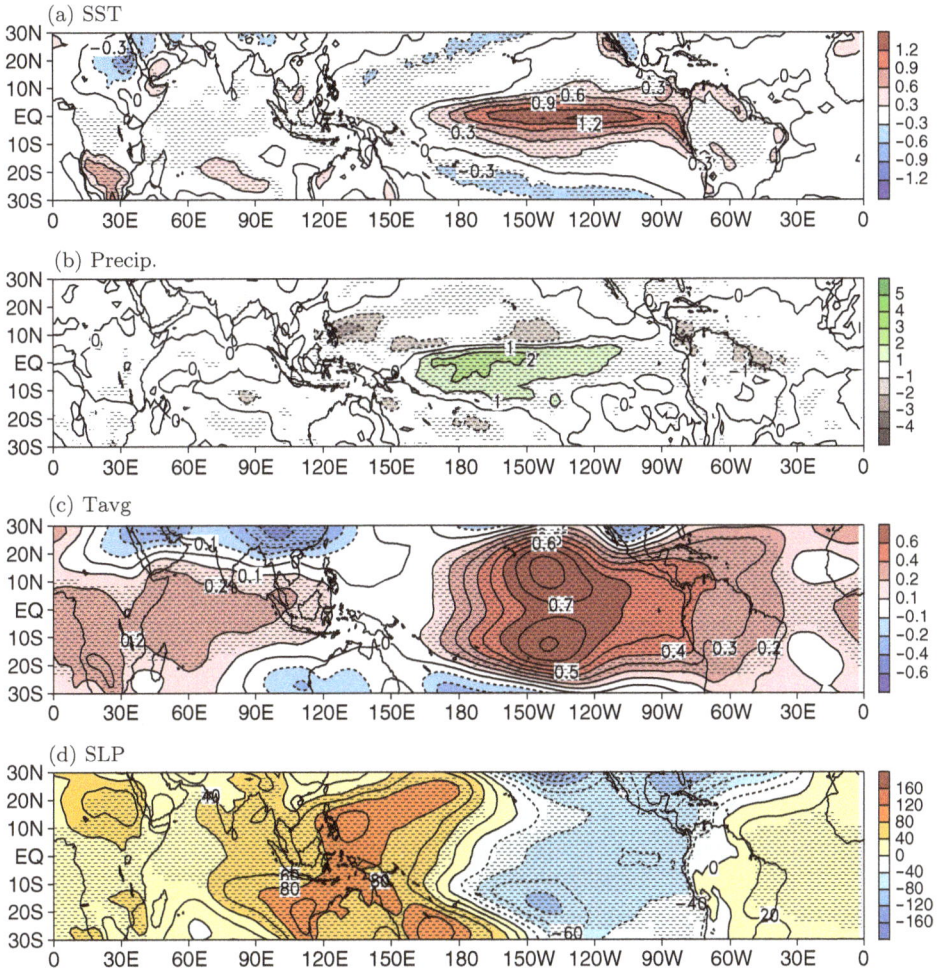

Figure 3.1. For December–February (DJF), regression onto the Niño–3.4 SST index of, (a) SST ($K°C^{-1}$), (b) precipitation ($mm\,day^{-1}\,°C^{-1}$), (c) tropospheric temperature ($K°C^{-1}$), and (d) SLP ($Pa°C^{-1}$). Data is from the NCEP–NCAR reanalysis for the period 1980–2008. A two-tailed t test was applied to the regression values and regions stippled correspond to the 99% confidence. (Figure 1 in [Ji et al., 2015] © American Meteorological Society. Used with permission.)

anomalies in vertical mean tropospheric temperature are positive over a broad region of the tropical central and eastern Pacific. Around the equator, the largest values of these anomalies are to the east of the largest precipitation anomalies corresponding to regions of deep convection (Fig. 3.1(c)). The magnitude of the tropospheric temperature anomalies drops off sharply from around the date line toward the western Pacific. The sea level pressure (SLP) anomalies are

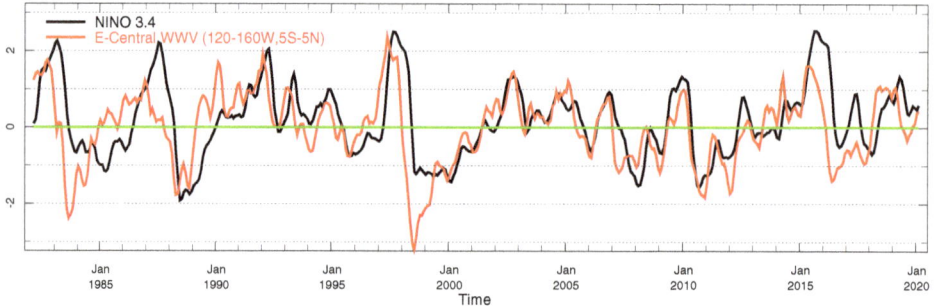

Figure 3.2.   Three-month running average time series of the standardized anomalous warm water volume in the east-central equatorial Pacific (red line), and the standardized Niño–3.4 anomalies from 1982 to 2020 (black line). (See https://iridl.ldeo.columbia.edu/maproom/ENSO/Time_Series/Heat_Storage_ECent_Pac.html)

reminiscent of the classic Southern Oscillation pattern: strong negative and positive anomalies in the eastern and western Pacific, respectively (Fig. 3.1(d)).

The black curve in Fig. 3.2 shows the time series of the Niño–3.4 SST anomalies expressed as the three-month running average of their standardized values. The red curve in Fig. 3.2 shows a similar time series for anomalies in warm water volume (WWV) in the east-central equatorial Pacific, expressed as the depth of water above the 20°C potential temperature isotherm averaged over the region (5°S–5°N, 160°W–120°W). The 20°C potential temperature isotherm depth anomalies are calculated with respect to the 1982–present base period. A low-pass filter was applied to the anomalies to remove the variability with timescales shorter than one year. According to Fig. 3.2, anomalous warm SST events (El Niño) occur approximately every 3–5 years. These events are accompanied by a deepening of the thermocline in the central-to-eastern equatorial Pacific, while cold events (La Niña) are accompanied by a shoaling of the thermocline in the region.

Much of the theoretical work on ENSO focuses on its oscillatory aspect associated with a dominant spectral peak. The mature El Niño and La Niña are viewed as the positive and negative peak phases of quasi-biennial oscillation in SST over the central and eastern tropical Pacific Ocean in association with a deepening and shoaling of the thermocline, respectively. These thermocline variations can also be expressed as variations in oceanic heat content (defined, for example, as the mean temperature in the upper ocean) or the WWV mentioned in the previous paragraph. The Bjerknes feedback among SST, wind, ocean currents, and subsurface temperature anomalies is active during these phases [Bjerknes, 1966, 1969]. The reigning paradigm for the transition phases of the oscillation is that subsurface ocean temperature anomalies carry the memory of the phenomenon when little anomaly may be visible at the surface. The conceptual models presented in the following two sections are different ways of interpreting the mechanisms that carry the memory. The presentation follows closely the discussion in Mechoso et al. [2003].

3.3   The recharge-oscillator model

The recharge-oscillator theory emphasizes the importance for the phase reversal of the ENSO cycle of the buildup (i.e., charge) and release (i.e., discharge) of ocean heat content in the equatorial band. The charge and discharge are generated by the non-equilibrium between the zonal-mean ocean heat content and wind stress at the equator. The terminology of a recharge-oscillator is a little questionable because the equation representing heat content change is a simplification of wave-dynamical processes and the process is linear, as opposed to the sudden capacitor-discharge-like process alluded to in the recharge-oscillator terminology. The JIN97 model simplifies the coupled system to two linear, ordinary differential equations. In this simplified scenario, the oscillation depends crucially on the time scale of equilibration in the SST equation in addition to the time scale of heat content build-up.

In JIN97, a positive SST anomaly in the onset of the peak (warm) stage induces a positive (westerly) wind stress anomaly at the equator. This wind anomaly has two effects. First, it induces a divergence of the oceanic Sverdrup transport at the equator that leads to a gradual reduction in zonal mean thermocline depth. Second, it sets up a flattening of the equatorial thermocline with a magnitude that is proportional to the wind stress. This leads to a deepening of the thermocline in the eastern part of the basin, which participates in the positive feedback with SST and wind stress described in the Bjerknes hypothesis, eventually bringing the oscillation to the peak phase in SST. As this ends the zonal-mean equatorial thermocline is shallowest. Climatological upwelling and the anomalies in subsurface temperature produced by shallow thermocline in the eastern part of the basin result in (i) a negative SST anomaly developing there, which induces a negative (easterly) wind stress anomaly, (ii) a deepening of the thermocline in the west, and (iii) a gradual increase of the mean thermocline depth (charge). This conceptual recharge-oscillator sets out to produce a simple set of equations that link anomalies in SST variations, thermocline depth adjustment, and wind stress at the equator.

JIN97 assumes that at the equator differences in thermocline depth anomaly and mean wind stress anomaly across the basin ($[\tau]$) are in approximate balance. This can be formulated as [Mechoso et al., 2000]

$$\frac{1}{\lambda K_E} H_E = \frac{1}{\lambda K_W} H_W + [\tau] \tag{3.1}$$

where H_E and H_W are the upper-ocean heat content anomalies in the equatorial eastern and western Pacific, respectively (defined as the average temperature anomaly above 300 m), and the coefficients K_E, K_W and λ are assumed to be constant.

The first of the two prognostic equations in the recharge-oscillator model synthesizes the equatorial adjustment

$$\frac{dH_W}{dt} = -rH_W - \alpha\,[\tau] \qquad (3.2)$$

where the first term in the right-hand side represents the overall effect of oceanic adjustment processes assumed to act at a constant rate r, and the second term represents the Sverdrup transport across the basin.

The second prognostic equation used in the recharge-oscillator model is based on the SST equation for a box in the eastern Pacific

$$\frac{dT_E}{dt} = -cT_E + \gamma H_E \qquad (3.3)$$

where (T_E) represents the SST anomalies in the eastern part of the basin. The first term in the right-hand side of Eq. (3.3) represents decay processes due to heat exchange between atmosphere and ocean acting at a constant rate c, and the second term represents thermocline feedback processes mediated by upwelling. Feedback processes associated with upwelling and zonal advection by surface layer currents can implicitly affect c, since they have little time lag relative to wind stress anomalies and thus to SST anomalies. Zonal advection by the geostrophic current can be incorporated into thermocline feedback process [An and Jin, 2001].

The coupling condition in JIN97 is the requirement that mean wind stress anomaly across the basin ($[\tau]$) and the SST anomalies in the eastern part of the basin (T_E) satisfy a linear relationship, where b is a constant,

$$[\tau] = bT_E \qquad (3.4)$$

Substitution of the coupling assumption Eq. (3.4) in Eq. (3.2) gives

$$\frac{dH_W}{dt} = -rH_W - \alpha bT_E \qquad (3.5)$$

Using Eqs. (3.1) and (3.4) in Eq. (3.3) we obtain

$$\frac{dT_E}{dt} = RT_E + K_E K_W^{-1}\gamma H_W \qquad (3.6)$$

where the parameter

$$R = K_E \lambda b\gamma - c \qquad (3.7)$$

represents the Bjerknes positive feedback process of tropical atmosphere-ocean interactions as it includes b, which is defined in the coupling condition Eq. (3.4). The oscillation equation results from writing H_W in Eq. (3.5) as a function of T_E using Eq. (3.6), which yields

$$\frac{d^2 T_E}{dt^2} - (R - r) \frac{d T_E}{dt} + \left(K_E K_W^{-1} ab\gamma - Rr \right) T_E = 0 \qquad (3.8)$$

where Eq. (3.1) was also used. Note that the combination of parameters $K_E K_W^{-1} ab\gamma - Rr$ is the leading effect tending to produce oscillatory behavior. The period of oscillatory solutions is given by

$$P = 2\pi \left[K_E K_W^{-1} ab\gamma - \left(R^2 + r^2 \right)/4 \right]^{-1/2} \qquad (3.9)$$

The magnitude of the parameters in Eq. (3.9) can be estimated on the basis of heuristic arguments. Mechoso et al. [2003] estimated these parameters according to the characteristics of the leading oscillatory mode in a realistic simulation of ENSO by a numerical climate model [Yu and Mechoso, 2001; see Ch. 7]. A fitting of the simulated results to the different equations above suggests the following set of parameters: $r = 0.108$ month^{-1}; $ab = 0.104$ month^{-1}; $R = 0.0988$ month^{-1}; and $K_E \gamma / K_W = 0.269$ month^{-1}. Using these values, the resulting period is 4.0 years with a slow decay value of 9 years^{-1}. Note that once these values are selected, then the constants for the physical processes are set up, and hypotheses made to estimate the values of these parameters do not necessarily hold. That is, a purely empirically motivated oscillator equation or second order auto-regressive process could be fitted to the period of the observation.

The coupling given by Eq. (3.4) implies a simple proportionality between the wind stress and T_E. However, it is reasonable to assume that the wind stress adjusts to SST with a finite adjustment time, which is shorter than the ENSO period.

For $T_E = \widehat{T_E} \exp(i\omega t)$, a modified coupling condition can be

$$[\tau] = b \widehat{T_E} \left(1 - i\omega\eta \right)^{-1} e^{i\omega t} \qquad (3.10)$$

which represents a frequency-dependent reduction of amplitude that damps higher frequencies combined with a lag of approximately η for $\omega\eta$ small.

Using a set of parameters that gives a period of 2.7 years for the JIN97 model with the coupling condition given by Eq. (3.4), the condition given by Eq. (3.10) yields a period of 4.2 yr. Thus, the atmospheric adjustment time, though small, can have a large effect [see Mechoso et al. 2003, for more details].

3.4   The delayed oscillator model

In the SSBH model, the delayed negative feedback associated with free oceanic equatorial wave propagation and reflection at the western boundary is assumed to be responsible for the phase reversal of the ENSO cycle. The model equations can be written in a form very closely related to the recharge-oscillator. The relationship for T_E also remains the same. However, the relationship for H_W is replaced by

$$H_W = -\alpha^*[\tau(t - \delta)] \tag{3.11}$$

where δ is the delay time, which can be interpreted as a weighted mean Rossby wave travel time; i.e., the distance from the central Pacific to the western boundary divided by the Rossby wave speed.

The coupling condition in SSBH is the same as in JIN97 (Eq. (3.4)). This gives

$$dT_E/dt = RT_E - K_E K_W^{-1} \gamma \alpha^* b T_E(t - \delta) \tag{3.12}$$

For oscillations of the form $\exp(i\omega t)$, the dispersion relation becomes

$$(i\omega - R) = -K_E K_W^{-1} \gamma \alpha^* b \exp[-i\omega\delta] \tag{3.13}$$

Using the same values R and $K_E\gamma/K_W$ above, and $\alpha^* b = 0.622$ $\delta = 6.0$ month, yields a period of 3.5 years and a decay rate of $(3.9\,\text{year})^{-1}$.

3.5   General comments on the recharge and delayed
oscillator models

As presented in the previous sections of this chapter, the simple models are treated in linear form. It is known from a variety of intermediate model studies [Battisti and Hirst, 1989; Jin and Neelin, 1993; Jin et al., 1996] that while the linear period captures the finite amplitude period to a first approximation, nonlinear processes can modify (often increase) the period. The period is much less sensitive than the growth rate, since the first effect of the nonlinearity is simply to balance the growth rate if the mode is unstable. If the mode is stable then the balance is between decay rate and stochastic input from the atmosphere.

In the phase plane of wind stress versus SST, the anomalies approximately follow a linear relationship adjustment time. From Syu and Neelin [2000] and Neelin et al. [2000], it is known that such an adjustment time, though short, can increase the period of ENSO models. This adjustment time can be long enough to suggest that it is not purely atmospheric, but rather involves ocean mixed layer processes outside the eastern equatorial Pacific, spinning up in conjunction with the atmosphere. Other adjustment processes, however, including coupled interactions between the atmosphere and the ocean surface mixed layer can easily modify the results. The need for additional physics for a better capturing of the oscillation period illustrates how additional mechanisms can affect the ENSO mode, for which there exists a number of observationally based indications [e.g., Wang et al., 1998; Weisberg et al., 1999]. The missing physics is reasonably accounted for by including an additional adjustment time between the wind stress and the eastern Pacific SST. This adjustment time can represent the coupled spin-up of the atmosphere and the ocean mixed-layer processes outside the equatorial eastern Pacific region on which both the recharge and delayed oscillator models focus.

Other hypotheses aimed to explain the oscillator nature of ENSO include the "western Pacific Oscillator" [Weisberg and Wang, 1997] and the "advective-reflective oscillator" [Picaut et al., 1997]. In the western Pacific oscillator, an interactive process between SST and sea level pressure over the off-equatorial western Pacific induces equatorial wind anomalies that affect ENSO. In the advective-reflective oscillator, the roles of oceanic wave reflection at both western and eastern boundaries are emphasized. These processes are posited to affect the anomalous zonal currents, which together with the mean current induce the zonal displacements of the western Pacific warm pool that characterize the ENSO oscillation. Basically, the delayed oscillator does not take into account neither the active air-sea coupling process in the western Pacific nor the wave reflection at the eastern Pacific. The recharge oscillator does not deal with oceanic wave processes either, but it represents them as intrinsic oceanic adjustment processes.

In Wang [2001a], all four oscillator models mentioned in this chapter are combined into a unified oscillator. This has four prognostic variables: (1) T, representing SST anomalies over the Niño3 region (5°S–5°N, 150°–90°W); (2) h, representing thermocline depth anomalies over off-equatorial northern western Pacific (8°–16°N, 140°–160°E); (3) τ_1, representing zonal wind stress anomalies over the Niño4 region (5°S–5°N, 160°E–150°W); and (4) τ_2, representing zonal wind stress anomalies over equatorial western Pacific (5°S–5°N, 120°–140°E). The unified oscillator equations are as follows:

$$\frac{dT}{dt} = a\tau_1 - b_1\tau_1(t - \eta) + b_2\tau_2(t - \delta) - \varepsilon T^3 \tag{3.14}$$

$$\frac{dh}{dt} = -c\tau_1(t - \lambda) - R_h h \tag{3.15}$$

$$\frac{d\tau_1}{dt} = dT - R_{\tau 1}\tau_1 \tag{3.16}$$

$$\frac{d\tau_2}{dt} = eh - R_{\tau 2}\tau_2 \tag{3.17}$$

where a, d, and e are feedback coefficients of Niño4 zonal wind stress anomalies to Niño3 SST anomalies, Niño3 SST anomalies to Niño4 zonal wind stress anomalies, and off-equatorial western Pacific thermocline anomalies to western Pacific zonal wind stress anomalies, respectively; ε, R_h, $R_{\tau 1}$, and $R_{\tau 2}$ are damping coefficients for each variable; b_1, b_2 and c are coefficients for delayed negative feedback processes through wave reflection at the western boundary, wind-forced wave contribution in the equatorial western Pacific, and ocean waves contribution by Niño4 zonal wind stress anomalies, respectively. If $b_2 = 0$ there is no contribution by ocean-atmosphere coupled processes in the western Pacific and the delayed oscillator is recovered. If $b_1 = 0$, there is no feedback due to equatorial Rossby wave reflection at the western boundary and the recharge oscillator is recovered. For recharge oscillator derivation, the steady state response of the atmosphere is assumed so that the time derivatives in Eqs. (3.16) and (3.17) can be set to zero. Since the recharge oscillator does not require the explicit role of wave propagation, all delay parameters can be set to zero. Then we have

$$\frac{dT}{dt} = \frac{ad - b_1 d}{R_{\tau 1}} T + \frac{b_2 e}{R_{\tau 2}} h - \varepsilon T^3 \qquad (3.18)$$

$$\frac{dh}{dt} = -\frac{cd}{R_{\tau 1}} T - R_h h \qquad (3.19)$$

These equations and those in Sec. 3.3 only differ in the cubic term. By adding a negative feedback term of wave reflection at the eastern boundary to the unified model, this can be transformed into the advective-reflective oscillator model. However, the original unified oscillator actually includes the advective-reflective oscillator implicitly. By adopting appropriate parameters, the unified model produces an interannual oscillation. Therefore, Wang [2001b] claimed that the oscillators mentioned above in this chapter can be considered special cases of the unified oscillator, and that in Nature all of the oscillatory mechanisms are operating.

3.6   Perspectives

Beyond the linear perspective on ENSO dynamics, several works have reported the positively skewed probability density distribution of ENSO indices [Deser and Wallace, 1987; Burgers and Stephenson, 1999; An, 2009] and the asymmetric evolution of the ENSO cycle [Larkin and Harrison 2002; An et al., 2005]. Phenomenologically, anomalies during El Niño are stronger than during La Niña [Deser and Wallace, 1987; Burgers and Stephenson, 1999; An, 2009]; El Niño is usually followed by La Niña but the opposite case can occur albeit less frequently; the termination of El Niño is usually faster than that of La Niña [Larkin and Harrison, 2002; Okumura and Deser, 2010]. Such asymmetric properties between El Niño and La Niña are integral part of ENSO complexity [Timmermann et al., 2018], and are attributed to either internal nonlinear dynamics or external impacts [e.g., Ohba and Ueda, 2007; Okumura et al., 2011]. This section mainly addresses internal nonlinear dynamics because external impacts are generally not incorporated into simple dynamical models.

The internal nonlinear processes in the tropical Pacific introduced in the scientific literature are as follows [see An et al., 2020 for more detailed descriptions]: (1) a stronger local air-sea coupling during El Niño compared to La Niña [Kang and Kug, 2002; Choi et al., 2013; Dommenget et al., 2013; Chen et al., 2016]; (2) nonlinear shortwave-cloud-SST interaction [Li and Philander, 1996; Li 1997; Lloyd et al., 2012]; (3) active westerly wind bursts during El Niño compared to La Niña [Jin et al., 2007; Levine and Jin, 2010]; (4) reduced bio-physical feedback associated with suppressed phytoplankton bloom during El Niño due to weakened upwelling [Marzeion et al., 2005; Timmermann and Jin, 2002]; (5) enhanced warming by nonlinear dynamical heating [An and Jin, 2004; Su et al., 2010]; (6) suppressed lateral mixing by tropical instability waves during El Niño [Yu and Liu, 2003; An, 2008]; (6) stronger upper ocean dynamic response to the wind stress anomalies during El Niño [Im et al., 2015; An and Kim, 2017, 2018]; (7) enhanced atmospheric noise activity including westerly wind events during El Niño [Eisenman et al., 2005; Jin et al., 2007; Kug et al., 2008; Hayashi and Watanabe, 2017]; etc.

To consider the different strength of local air-sea coupling during El Niño compared to La Niña, Choi et al. [2013] modified the delayed oscillator equation to show that an asymmetrical response in the intensity of the equatorial central Pacific wind stress anomaly to the equatorial eastern Pacific SST anomaly could lead to the asymmetrical amplitude/transition/evolution of El Niño and La Niña. In the delayed oscillator model, the asymmetric wind response to SST anomalies was taken into account by simply converting $T_E(t)$ to $T_E(t) + \rho|T_E(t)|$ such that

$$\frac{\partial T_E}{\partial t} = c[T_E(t) + \rho|T_E(t)|] - b[T_E(t-\delta) + \rho|T_E(t-\delta)|] \qquad (3.20)$$

where ρ indicates "asymmetricity" and thus the feedback for positive $T_E(t)$ is enhanced by a factor of ρ compared to that for negative $T_E(t)$. Choi et al. [2013] showed that an asymmetric transition behavior between El Niño and La Niña similar to the one observed was simulated by using the modified delayed oscillator in Eq. (3.20). The physical meaning in the simple conversion used in Choi et al. [2013] refers to the asymmetric wind response to SST anomalies, but it is applicable to any of the above mentioned nonlinear processes. In An and Kim [2017], the asymmetricity ρ is determined by asymmetry in the wind stress response to SST anomalies, asymmetry in upper ocean wave response to

the wind stress, or asymmetry in the subsurface temperature response to the
ocean wave change between El Niño and La Niña.

Another important update over earlier simple ENSO models is the incorporation of nonlinear processes in the linear recharge oscillator model [Kim and An,
2020; Jin et al., 2020]. Nonlinear dynamic heating is one of the main drivers for
a positively skewed ENSO intensity by enhancing the warming tendency during
El Niño and suppressing the cooling tendency during La Niña [An and Jin,
2004; Su et al., 2010]. As in Kim and An [2020], nonlinear dynamic heating was
approximated by the quadratic terms comprising (T_E, H_E), $\beta_1 T_E^2 + \beta_2 T_E H_E$.
In addition to nonlinear dynamic heating, the state-dependent noise forcing
can be included in the linear recharge oscillator model through terms of the
form $\sigma(1 + BH(T_E))\xi_t$, where $\sigma \xi_t$ is a stochastic noise with variance σ^2, B is a
positive constant, and $H(x)$ is a Heaviside function [e.g., Levine and Jin, 2010].
On the basis of such modifications, the linear recharge oscillator model becomes
a nonlinear recharge oscillator model as follows:

$$\frac{dT_E}{dt} = RT_E + \gamma H_E + \beta_1 T_E^2 + \beta_2 T_E H_E + \sigma(1 + BH(T_E))\xi_t \qquad (3.21)$$

Eq. (3.21) combined with Eq. (3.5) drives the positively skewed probability
density of ENSO (i.e. T_E) and asymmetric evolution of ENSO cycle with
$\beta_1 > 0$ or $B > 0$ [e.g., Kim and An, 2020]. This is because $\beta_1 T_E^2$ ($\beta_1 > 0$)
is always positive regardless of the sign of T_E — like nonlinear dynamic heating
would be — and the additional noise forcing by BT_E only boosts up El Niño.
Consequently, El Niño is enhanced while La Niña is suppressed.

Chapter 4

Models of Intermediate Complexity and ENSO Prediction

4.1 Introduction

The present chapter examines intermediate-complexity models of the atmosphere-ocean system. In this important model category both the atmosphere and ocean components are based on drastically simplified versions of dynamical and physical processes. These models are capable of making ensemble climate predictions months in advance with modest computational resources. Such a feature makes them also useful for parameter-sweep studies aimed at understanding the relative importance of fundamental physical processes controlling a complex climate phenomenon. A major application area for intermediate-complexity models is El Niño-Southern Oscillation (ENSO) — one of the most outstanding phenomena of the atmosphere-ocean system on Earth.

The oceanic manifestation of ENSO is an interannual oscillation in sea surface temperature (SST) that occurs quasi-regularly over the tropical Pacific. The atmospheric manifestation is a seesaw in surface pressure with centers over the eastern tropical Pacific and northern Australia (see Ch. 1). ENSO impacts can reach all over the globe and significantly affect climate variability in a wide range of space and time scales [Ropelewski and Halpert, 1986, 1987, 1989; Halpert and Ropelewski, 1992]. In view of this importance, intense efforts have been made to develop models capable of providing realistic simulations and skillful predictions of ENSO. In this regard, the success of intermediate-complexity models is widely acknowledged.

The presentation that follows summarizes the physical background, formulations, and application to forecasting of intermediate-complexity models. Most attention is dedicated to the one developed by Cane et al. [1986] and Zebiak and Cane [1987] (hereafter, the ZC87 model). ZC87 — which produced the first successful prediction of El Niño — has been widely used in fundamental studies as well as in predictions of ENSO. Other intermediate-complexity components that use statistical atmospheric components are also mentioned.

4.2   Gill-type model of the tropical atmosphere

Gill's (1980) model has been very influential in studies of the tropical atmosphere and most intermediate-complexity models adopt it to represent the atmosphere. The main energy source for the tropical atmospheric circulation is latent heating. Moisture condensation accompanied by deep convection occurs mainly near the mid-troposphere (400–600 hPa), where latent heat release is maximized. The vertical structure of the atmospheric response to such heating can be approximated by the first baroclinic mode, in which motions in the upper and lower troposphere are opposite to each other (e.g., convergence at upper levels and divergence at lower levels). Such a structure justifies the design of a model of the tropical atmosphere for simulating the first baroclinic mode under homogeneous, incompressible, and hydrostatic conditions.

We start by formulating the momentum, continuity and thermodynamic equations for a two-level atmosphere. Figure 4.1 shows the vertical structure of such model in which the horizontal velocity components $(\mathbf{v}_{i=1.3})$ and geopotential height $(\varphi_{i=1,3})$ are defined at the levels, and vertical velocities $\omega_{0,2,4}$ are defined at mid-level as well as at the surface and top where they are set to zero. Using the pressure coordinate, the momentum and continuity equations for level 1(upper) and level 3 (lower) are

$$\frac{D\mathbf{v}_1}{Dt} + f\mathbf{k} \times \mathbf{v}_1 = -\nabla\varphi_1 - A\mathbf{v}_1 \qquad (4.1a)$$

$$\nabla \cdot \mathbf{v}_1 + \frac{\omega_2 - \omega_0}{\Delta p} = 0 \qquad (4.1b)$$

$$\frac{D\mathbf{v}_3}{dt} + f\mathbf{k} \times \mathbf{v}_3 = -\nabla\varphi_3 - A\mathbf{v}_3 \qquad (4.1c)$$

$$\nabla \cdot \mathbf{v}_3 + \frac{\omega_4 - \omega_2}{\Delta p} = 0 \qquad (4.1d)$$

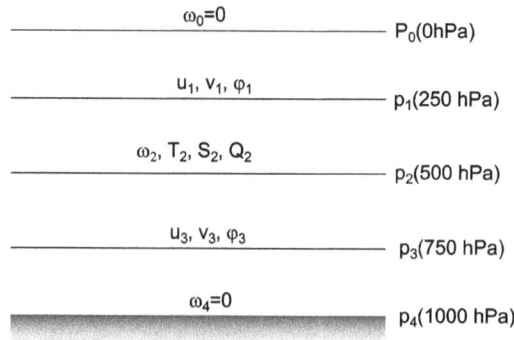

Figure 4.1.   Schematic diagram of the vertical structure and principal variables for a two-level atmosphere model: $\mathbf{v}_{i=1.3}$ indicate horizontal velocities, $\varphi_{i=1,3}$ is geopotential height, $\omega_{0,2,4}$ is vertical velocity, Q_2 is diabatic heating, and S_2 is static stability.

where f is the Coriolis parameter, A is damping rate and $\Delta p(\approx 500\,\text{hPa})$ is pressure depth of the half-atmosphere.

The thermodynamic equation at level 2 is

$$\frac{dT_2}{dt} - S_p \omega_2 = \dot{Q}_2/C_p \tag{4.2}$$

where S_p is the static stability of dry air,

$$S_p = -\bar{T}(\partial \ln \bar{\theta}/\partial p) \tag{4.3}$$

At the mid-troposphere $S_2 \approx 5 \times 10^{-4}\,\text{KPa}^{-1}$ [Holton, 2012]. The second term on the left-hand side of Eq. (4.2) represents adiabatic cooling through vertical motion, and the term on the right-hand side is diabatic heating by moisture condensation (Q). In the hydrostatic approximation ($\partial\varphi/\partial p = -\alpha$); where α is specific volume and using the ideal gas law ($p\alpha = R_d T$), T_2 is converted to the height difference between level 1 and level 3, such that $T_2 = -(p_2/R_d)[(\varphi_3 - \varphi_1)/\Delta p]$. Then, the thermodynamic equation becomes

$$\frac{D\varphi}{Dt} + \frac{R_d \Delta p}{p_2} S_p \omega_2 = -\frac{R_d \Delta p}{C_p p_2} \dot{Q}_2 \tag{4.4}$$

The equations for the first baroclinic mode are obtained by computing the difference between level 3 (Eq. (4.1c) and Eq. (4.1d)) and level 1 (Eq. (4.1a) and Eq. (4.1b)), and using the thermodynamic equation at the mid-level to eliminate ω_2:

$$\frac{D\mathbf{v}}{Dt} + f\mathbf{k} \times \mathbf{v} = -\nabla\varphi - A\mathbf{v} \tag{4.5a}$$

$$\frac{D\varphi}{Dt} + C_a^2 \nabla \cdot \mathbf{v} = -\lambda \dot{Q}_2 \tag{4.5b}$$

where $(\mathbf{v}, \varphi) = (\mathbf{v}_3 - \mathbf{v}_1, \varphi_3 - \varphi_1)/2$, $C_a^2 = (R_d \Delta p^2/2p_2)S_p \cong (60\,\text{ms}^{-1})^2$, and $\lambda = (R_d/2C_p)(\Delta p/p_2)$, in which C_a is the gravity wave speed corresponding to the first baroclinic mode of the tropical atmosphere. The homogeneous solution of Eq. (4.5a) and Eq. (4.5b) without forcing and the inhomogeneous solution with forcing are well-documented in An et al. [2021; Ch. 2].

Based on these expressions for the first baroclinic mode, Gill [1980] developed a model of the steady tropical atmosphere forced by convective heating. The steady state approximation is based on assuming a deep vertical structure (corresponding to the first baroclinic mode), a main heat source of convective heating in the mid-troposphere, strong damping that mediates the steady response, longwave approximation in the zonal direction, and the equatorial β-plane approximation ($f \approx \beta y$). The equations of Gill's model in

Cartesian coordinates are [see also An et al., 2021]

$$\epsilon u - \beta y v + \varphi_x = 0 \tag{4.6a}$$

$$\epsilon v + \beta y u + \varphi_y = 0 \tag{4.6b}$$

$$\epsilon \phi + c^2(u_x + v_y) = -Q \tag{4.6c}$$

where c is a characteristic speed that depends on atmospheric static stability, ϵ is a coefficient of linear friction/Newtonian damping; and Q is diabatic heating. Possible values for these parameters are $c = 45 \, \mathrm{ms}^{-1}$ and $\epsilon = (2.5 \, \mathrm{days})^{-1}$.

4.3   One- and two-layer shallow water systems

These systems consist of either a single layer or two layers of fluid, respectively, with different density on a rotating plane (see Figs. 4.2 and 4.3). Motions are assumed to be independent of height and vertical variations of pressure are described by the hydrostatic approximation. Such layered systems provide the foundation for the oceanic component in models of intermediate complexity.

4.3.1   One-layer shallow-water system

4.3.1.1   Governing equations

Because horizontal velocities are independent of height (z), the horizontal momentum equation can be written as follows:

$$\frac{D\mathbf{v}}{Dt} + f\mathbf{k} \times \mathbf{v} = -\frac{1}{\rho}\nabla P \tag{4.7}$$

where $\mathbf{v}$ is velocity $(\mathbf{v} = \mathbf{v}(x, y, t))$, $f = 2\Omega$, P is pressure with reference to the atmospheric pressure, the symbol ∇ denotes the horizontal gradient operator,

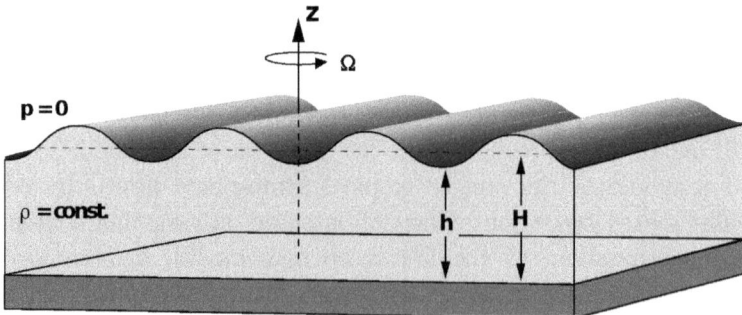

Figure 4.2.   Schematic diagram of the one-layer shallow water system.

ρ is density, g is gravity, and Ω is angular rotation speed. Moreover,

$$\frac{D\mathbf{v}}{Dt} = \frac{\partial \mathbf{v}}{\partial t} + \nabla \mathbf{v} \cdot \mathbf{v} \tag{4.8}$$

The left-hand side of Eq. (4.7) is independent of the vertical coordinate. The driving of a motion that is independent of height implies a pressure gradient $(-\nabla P)$ that is also independent of height. Vertical variations of pressure are given by the hydrostatic relationship:

$$P = \rho g(h - z) \tag{4.9}$$

where h is the height of the free surface. Therefore,

$$-\frac{1}{\rho}\nabla P = g\nabla h \tag{4.10}$$

Using Eq. (4.10) in Eq. (4.7) provides a relationship between the horizontal velocity and the height of the free surface:

$$\frac{D\mathbf{v}}{Dt} + f\mathbf{k} \times \mathbf{v} = -g\nabla h \tag{4.11}$$

Also, from Eq. (4.8),

$$\frac{\partial \mathbf{v}}{\partial t} = -\nabla \mathbf{v} \cdot \mathbf{v} - f\mathbf{k} \times \mathbf{v} - g\nabla h \tag{4.12}$$

This latter equation excludes vertically varying, purely inertial motions. The continuity equation for an incompressible fluid is

$$\frac{\partial w}{\partial z} = -\nabla \cdot \mathbf{v} \tag{4.13}$$

The left-hand side of Eq. (4.13) is independent of height; therefore, the vertical velocity must be linear in height. If the lower boundary is horizontal $(z = 0)$, then $w(x, y, 0, t) = 0$, and

$$\frac{\partial w}{\partial z} = \frac{1}{h}w_{z=h} = \frac{1}{h}\frac{Dh}{Dt} \tag{4.14}$$

where $w_{z=h}$ is the vertical velocity at the free surface. Thus, Eq. (4.13) becomes

$$\frac{Dh}{Dt} = -h\nabla \cdot \mathbf{v} \tag{4.15}$$

or,

$$\frac{\partial h}{\partial t} = -\nabla \cdot (h\mathbf{v}) \tag{4.16}$$

According to Eq. (4.12) and Eq. (4.16), the motions of a homogeneous layer of fluid that satisfies the quasi-static approximation are governed by

$$\frac{\partial u}{\partial t} + u\frac{\partial u}{\partial x} + v\frac{\partial u}{\partial y} - fv = -g\frac{\partial h}{\partial x} \tag{4.17a}$$

$$\frac{\partial v}{\partial t} + u\frac{\partial v}{\partial x} + v\frac{\partial v}{\partial y} + fu = -g\frac{\partial h}{\partial y} \tag{4.17b}$$

$$\frac{\partial h}{\partial t} + \frac{\partial (hu)}{\partial x} + \frac{\partial (hv)}{\partial y} = 0 \tag{4.17c}$$

4.3.1.2 Energy equation

To obtain an expression for the kinetic energy variation in the one-layer shallow water system, Eq. (4.11) is multiplied by $h\rho\mathbf{v}$

$$h\frac{D}{Dt}\left(\frac{1}{2}\rho\mathbf{v}^2\right) = -\rho g\nabla\left(\frac{h^2}{2}\right)\cdot\mathbf{v} \tag{4.18}$$

This can also be written as

$$\frac{D}{Dt}\left(\frac{1}{2}\rho\mathbf{v}^2 h\right) = \frac{1}{2}\rho\mathbf{v}^2\frac{Dh}{Dt} - \rho g\nabla\left(\frac{h^2}{2}\right)\cdot\mathbf{v} \tag{4.19}$$

The bracketed term in the left-hand side of Eq. (4.19) is the kinetic energy (K_c) of the fluid column:

$$K_c = \int_0^h \frac{1}{2}\rho\mathbf{v}^2 dz = \frac{1}{2}\rho h\mathbf{v}^2 \tag{4.20}$$

Hence, Eq. (4.19) becomes

$$\frac{DK_c}{Dt} = \nabla K_c \cdot \mathbf{v} - \nabla\cdot(K_c\mathbf{v}) - \rho g\nabla\left(\frac{h^2}{2}\right)\cdot\mathbf{v} \tag{4.21}$$

Similarly, for the potential energy, multiplication of Eq. (4.15) by ρgh yields

$$\frac{D}{Dt}\left(\frac{1}{2}\rho gh^2\right) = -\rho gh^2\nabla\cdot\mathbf{v} \tag{4.22}$$

which can be written as

$$\frac{D}{Dt}\left(\frac{1}{2}\rho gh^2\right) = -\rho g\frac{h^2}{2}\nabla\cdot\mathbf{v} - \rho g\frac{h^2}{2}\nabla\cdot\mathbf{v}$$

$$= \nabla\left(\frac{1}{2}\rho gh^2\right)\cdot\mathbf{v} - \nabla\cdot\left(\frac{1}{2}\rho gh^2\mathbf{v}\right) - \rho g\frac{h^2}{2}\nabla\cdot\mathbf{v} \tag{4.23}$$

The bracketed term in the left-hand side of Eq. (4.23) is the potential energy (P_c) of the fluid column,

$$P_c = \int_0^h \rho g z \, dz = \frac{1}{2} \rho g h^2 \qquad (4.24)$$

while the right-hand side of the same expression represents the conversion between kinetic and potential energies. Equation (4.23) becomes

$$\frac{DP_c}{Dt} = \nabla P_c \cdot \mathbf{v} - \nabla \cdot (P_c \mathbf{v}) - \rho g \left(\frac{h^2}{2} \right) \nabla \cdot \mathbf{v} \qquad (4.25)$$

Combining Eq. (4.21) and Eq. (4.25) yields

$$\frac{DK_c}{Dt} + \frac{DP_c}{Dt} = \nabla K_c \cdot \mathbf{v} + \nabla P_c \cdot \mathbf{v} - \nabla \cdot (K_c \mathbf{v}) - \nabla \cdot (P_c \mathbf{v}) - \nabla \left(\rho g \frac{h^2}{2} \mathbf{v} \right) \qquad (4.26)$$

To interpret the last term on the right-hand side of Eq. (4.26) we compute

$$\int_0^h p\mathbf{v} \, dz = \int_0^h \rho g (h - z) \mathbf{v} \, dz = \rho g \mathbf{v} \int_0^h (h - z) dz = \rho \frac{1}{2} g h^2 \mathbf{v} \qquad (4.27)$$

Replacing in Eq. (4.26) gives the energy equation for a fluid column:

$$\frac{\partial K_c}{\partial t} + \frac{\partial P_c}{\partial t} = -\nabla \cdot (K_c \mathbf{v}) - \nabla \cdot (P_c \mathbf{v}) - \nabla \cdot \left(\int_0^h p\mathbf{v} \, dz \right) \qquad (4.28)$$

Let overbars denote the horizontal-area average over an infinite domain or a domain with boundaries in which the motion is either periodic or the normal component of $\mathbf{v}$ vanishes (material boundary). In these cases, we can easily see from Eq. (4.22) that $\partial(\overline{K_c + P_c})/\partial t = 0$.

Next, let $H = \bar{h}$, and $h' = h - H$, where H is a constant. From Eq. (4.24), we can compute the mean potential energy in the domain

$$\overline{g \frac{1}{2} h^2} = \overline{g \frac{1}{2} (H + h')^2} = g \left(\frac{1}{2} H^2 + \frac{1}{2} \overline{h'^2} \right) \qquad (4.29)$$

Moreover,

$$\frac{\partial}{\partial t} \overline{\rho g \frac{1}{2} h^2} = \frac{\partial}{\partial t} \overline{\rho g \frac{1}{2} h'^2} \qquad (4.30)$$

The right-hand side of Eq. (4.30) is the (mean) *available potential energy* of the system. It represents the excess of the potential energy over its minimum value.

4.3.1.3 *Vorticity equation*

The expression for the vertical component of the relative vorticity, $\zeta = \mathbf{k} \cdot \nabla \times \mathbf{v}$ is, according to Eq. (4.12),

$$\frac{\partial \zeta}{\partial t} = -\mathbf{v} \cdot \nabla (f + \zeta) - (f + \zeta)\nabla \cdot \mathbf{v} \qquad (4.31)$$

Eliminating $\nabla \cdot \mathbf{v}$ between Eq. (4.15) and Eq. (4.31) yields

$$\frac{\partial}{\partial t}\left(\frac{f + \zeta}{h}\right) = -\mathbf{v} \cdot \nabla \left(\frac{f + \zeta}{h}\right) \qquad (4.32)$$

The bracketed term on the left-hand side of Eq. (4.32) is the *potential vorticity* of the system. Thus,

$$\frac{D}{Dt}\left(\frac{f + \zeta}{h}\right) = 0 \qquad (4.33)$$

which means that the fluid columns conserve their potential vorticity as they move (in the absence of dissipation).

4.3.2 **Two-layer shallow-water system (linearized)**

Let us consider a two-layer shallow water system open to the atmosphere in which ρ_1, ρ_2 ($\rho_2 > \rho_1$) are densities of the upper and lower layers and H_1, H_2 are mean depths of the layers (see Fig. 4.3). In this subsection, we will assume that all velocities are small, and that perturbations in the depth of the layers are much smaller than the mean depths. We also add terms representing stresses at the upper, interface, and bottom surfaces ($\vec{\tau}_s, \vec{\tau}_I, \vec{\tau}_b$, respectively). The linearized

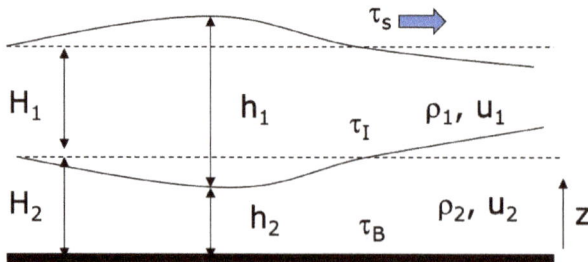

Figure 4.3.   Schematic diagram for the two-layer shallow-model.

form of Eq. (4.11) and Eq. (4.15) govern the motions of each layer:

$$\frac{\partial \mathbf{v}_1}{\partial t} + f\mathbf{k} \times \mathbf{v}_1 = -\frac{1}{\rho_1}\nabla P_1 + \frac{\vec{\tau}_s - \vec{\tau}_I}{\rho_1 H_1} \tag{4.34a}$$

$$\frac{\partial h_1}{\partial t} = -H_1 \nabla \cdot \mathbf{v}_1 \tag{4.34b}$$

$$\frac{\partial \mathbf{v}_2}{\partial t} + f\mathbf{k} \times \mathbf{v}_2 = -\frac{1}{\rho_2}\nabla P_2 + \frac{\vec{\tau}_I - \vec{\tau}_B}{\rho_2 H_2}, \tag{4.34c}$$

$$\frac{\partial h_2}{\partial t} = -H_2 \nabla \cdot \mathbf{v}_2 \tag{4.34d}$$

Pressures at depth z_1 in layer 1 and z_2 in layer 2, $P_1(z_1)$ and $P_2(z_2)$, are determined by the hydrostatic equation, such that $P_1(z_1) = \rho_1 g(h_1 + h_2 - z_1)$ and $P_2(z_2) = \rho_1 g h_1 + \rho_2 g(h_2 - z_2)$. Then, the pressure gradients terms in each layer are

$$\frac{1}{\rho_1}\nabla P_1 = g\nabla(h_1 + h_2) \tag{4.35a}$$

$$\frac{1}{\rho_2}\nabla P_2 = g\nabla(h_1 + h_2) - g\delta\nabla h_1, \qquad \delta = \frac{\rho_2 - \rho_1}{\rho_2} \ll 1 \tag{4.35b}$$

The term $g\delta$ is referred to as reduced gravity: $g'(\sim 1 - 2\,\mathrm{cm s}^{-2})$. In this term gravity is scaled by the normalized density difference between layers, which depends on the ocean stratification (or static stability). Stronger ocean stratification (i.e., warmer upper layer and cooler lower layer) implies a higher value of reduced gravity.

4.3.3 *Reduced gravity ocean model*

The two-layer shallow water systems described in Sec. 4.3.1 and Sec. 4.3.2 are widely used in ocean applications. In the tropical oceans, water density tends to increase in depth albeit not at a uniform rate. A well-mixed layer of approximately constant density is found just below the ocean surface (see Ch. 1). This special layer, which is referred to as the "surface mixed layer", has its counterpart in the atmospheric PBL (see Ch. 5). Below about 300 m in depth, i.e., in the "deep ocean", the water density is also approximately constant in depth. In between the surface mixed layer and the deep ocean there is a prominent thermocline where the vertical temperature/density gradient is steep (see Fig. 1.3). This special configuration motivates a conceptual representation of the ocean as a two-layer fluid system with different densities (lighter on top), and in which the thermocline is collapsed to a surface where steep vertical gradients are concentrated. (Note that a similar approach is followed with the mixed layer in the atmospheric PBL, see Ch. 5.) Additional kinematic conditions include assuming that the motion in the deep ocean (lower layer) is much slower than that in the upper ocean (mixed layer). This is a plausible assumption if the lower layer is much deeper (and thus is much more massive) than the upper layer. The special configuration of the two-layer model with only one active layer is referred to as a reduced gravity model. Such a configuration is especially applicable to the study of motions in which the first baroclinic mode in the vertical plays a key role.

If the upper layer is much shallower than the lower layer, i.e., $H_2 \gg H_1$, then the last term on the right-hand side of Eq. (4.34c) is very small and the lower layer can be assumed to be motionless, i.e., $\mathbf{v}_2 \approx 0$ and $\nabla P_2 \approx 0$.

The very weak horizontal temperature gradients — shown in Fig. (4.4) below $300\,\text{m}$ depth over the tropical Pacific — provide the justification for this approximation. From Eq. (4.35b), we have

$$\nabla(h_1 + h_2) \approx \delta \nabla h_1 \tag{4.36}$$

Finally, the linearized equations for a reduced-gravity model consisting of one active shallow-water layer on top of a very deep layer, and dropping the subscript "1" from the perturbation variables, are

$$\frac{\partial \mathbf{v}}{\partial t} + f\mathbf{k} \times \mathbf{v} = -g'\nabla h + \frac{\vec{\tau}_s}{\rho H} \tag{4.37a}$$

$$\frac{\partial h}{\partial t} + H\nabla \cdot \mathbf{v} = 0. \tag{4.37b}$$

where $\vec{\tau}_s$ represents a body force on the upper layer. Also, from Eq. (4.35a) and Eq. (4.36),

$$\nabla p = \delta \, \nabla h \tag{4.38}$$

where $p = P/(\rho g)$.

The reduced gravity model is formally equivalent to the one-layer shallow water model (Sec. 4.3.1), except for the modified gravity parameter. Physically, the reduced gravity model represents the baroclinic component of motion, whereas the shallow water model represents the barotropic component. Equation (4.31) was derived from the two-layer model; thus, it represents the first or gravest baroclinic mode. However, by simply changing the reduced gravity wave speed, it can be applied to the higher order baroclinic modes. The reduced gravity wave speed is inversely proportional to the order of the mode [Dewitte et al., 2003].

In the tropics, fast-moving long waves play a primary role in the ocean adjustment to variations in forcing as compared with the slow-moving short waves. In this context, the long-wave approximation ($L_x \gg L_y$) can be applied for filtering out unnecessary small-scale waves, including gravity, inertial gravity, and short Rossby waves. The meridional acceleration can also be neglected because the meridional scale of motions is small compared with the zonal scale ($L_y \approx \varepsilon$). Consequently, the equations for upper ocean dynamic under the long-wave approximation become

$$\frac{\partial u}{\partial t} - \beta y v = -g'\frac{\partial h}{\partial x} + \frac{\tau_s^x}{\rho H} \tag{4.39a}$$

$$\beta y u = -g'\frac{\partial h}{\partial y} + \frac{\tau_s^y}{\rho H} \tag{4.39b}$$

$$\frac{\partial h}{\partial t} + H\left(\frac{\partial u}{\partial x} + \frac{\partial v}{\partial y}\right) = 0 \tag{4.39c}$$

In several ENSO studies [e.g., McCreary, 1983], the depth h is replaced by $(H+p/g')$ and the characteristic speed of the first baroclinic mode $c = (g'H)^{1/2}$ is introduced in the formulation. Using these notations, the equations for the reduced gravity ocean model are

$$\frac{\partial u}{\partial t} - \beta y v + \frac{\partial p}{\partial x} = \frac{\tau_s^x}{\rho H} \tag{4.40a}$$

$$\frac{\partial v}{\partial t} + \beta y u + \frac{\partial p}{\partial y} = \frac{\tau_s^y}{\rho H} \tag{4.40b}$$

$$\frac{1}{c^2}\frac{\partial p}{\partial t} + \frac{\partial u}{\partial x} + \frac{\partial v}{\partial y} = 0 \tag{4.40c}$$

4.4 Oscillations in a reduced gravity ocean model coupled to a conceptual atmosphere

In the following sections we move to couple the different atmospheric and oceanic models reviewed so far in this chapter. We start with a very simple framework: A reduced gravity ocean model that is coupled to a conceptual atmosphere [McCreary, 1983; McCreary and Anderson, 1984]. Despite the simplicity, this framework gave deep insight into the oscillatory feature of ENSO in the early 1980s. In fact, the atmospheric model was drastically reduced to the coupling conditions in which the wind stress was explicitly written in terms of thermocline depth (h) representing SST in the following way,

$$\tau^x = \begin{cases} \tau_w + \tau_s, & \text{if } h_e < h_c \\ \tau_s, & \text{if } h_e \geq h_c \end{cases} \tag{4.41}$$

where h_e represents thermocline depth in the eastern part of the domain and h_c is a critical value for this depth. The zonal wind stress distributions were defined by analytic expressions representing a perturbation component $\tau_w = \tau_{0w}F(x,y)$, indicating the Walker circulation, superimposed on a seasonally varying component $\tau_s = \tau_{0s}F(x,y)\cos(\sigma t)$, where σ is the annual frequency. τ_{0w} and τ_{0s}

are $-0.05\,\mathrm{Nm^{-2}}$ and $0.015\,\mathrm{Nm^{-2}}$, respectively (McCreary and Anderson 1984). The pattern of $F(x,y)$ is zonally elongated along the equator with maximum values in the central Pacific mimicking the trade wind distribution (see details on these structures in McCreary and Anderson 1984).

The system so defined produced a turnabout feature between the warm and cold conditions in a certain range of h_c. Let us start with $\tau_s = 0$, i.e., the wind stress is constant in time. In this case, the system has two equilibrium states according to whether τ_w is turned on or off. When such system is perturbed, it tends to remain around one of the equilibrium states depending on the initial conditions. However, if τ_s represents a zonal wind distribution with a nonzero seasonal cycle, the system never reaches either equilibrium state because τ_w turns off or on. A shift from one equilibrium state to the other occurs as follows: when τ_w is turned on, upwelling equatorial Kelvin waves propagate into the eastern ocean where the thermocline shoals and h_e decreases. These upwelling Kelvin waves reflect at the eastern boundary as upwelling Rossby waves, which are eventually reflected at the western boundary as upwelling Kelvin waves. τ_w also forces downwelling Rossby waves, which eventually reflect from the western boundary as downwelling reflected Kelvin waves. These reflected Kelvin waves contribute to increase the depth h_e. After several reflections at the eastern and western boundaries until relaxing close to the equilibrium value of h_e corresponding to τ_w and additional contributions from τ_s, the condition $h_e \geq h_c$ is met and τ_w is turned off. At that time, h_e starts relaxing toward the other equilibrium state corresponding to $\tau_w = 0$. In such a model framework, ENSO is viewed as a series of events rather than a cyclic phenomenon [e.g., Kessler, 2002].

Earlier studies also provided a concept of oceanic adjustment to a given wind stress forcing. Highlighting of this concept would turn into an important contribution to the development of conceptual frameworks that can mimic oscillatory mechanisms, such as the delayed and recharge oscillator paradigms (Secs. 3.3 and 3.4 in this book). However, an excessively simplified formulation for the atmosphere model might not yield a realistic ENSO feature in the simulated SST.

4.5 Zebiak and Cane model (ZC87)

ZC87 [Zebiak and Cane, 1987] is a coupled atmosphere-ocean model specifically designed for ENSO simulation and prediction. As such, its emphasis is on anomalies from a seasonally-varying climatology and the model domain is confined over the tropical Pacific. Climatological mean states are prescribed from the observation. The atmosphere component in the ZC87 model design was based on Gill's [1980] model introduced in Sec. 4.2, and the ocean component followed the reduced gravity model formulation presented in Sec. 4.3.3. In this section, we describe these two components and their coupling in the ZC87 model. More detailed information is in Zebiak [1984, 1986] and in Zebiak and Cane [1987].

4.5.1 *Atmospheric component*

Of primary importance in Gill's [1980] type of models is the expression for the heating anomaly Q representing the latent heat release in the mid troposphere. Moisture is supplied by evaporation at the surface and convergence in the lower atmosphere. As the SST changes, the saturation vapor pressure and evaporation from the ocean surface also change. For convenience in the notation, we will represent the SST by T in the equations of this subsection. The relationship between the ocean temperature and saturation vapor pressure at the surface e_s and T is determined by the Clausius–Clapeyron equation:

$$\frac{de_s}{dT} = \frac{L_c}{R_v} \frac{e_s}{T^2}$$

(4.42)

where L_c and R_v are the specific heat of evaporation of water $(2.45 \times 10^6 \, \mathrm{J\,kg^{-1}})$ and the gas constant for water vapor $(461.5 \, \mathrm{J\,kg^{-1}K^{-1}})$, respectively. Using the Clausius–Clapeyron equation, the change in saturation vapor pressure associated with a change in ocean temperature (ΔT) about a reference temperature $(T_{ref} \approx 303 \, \mathrm{K})$ can be estimated as

$$\Delta e_s \cong \left. \frac{de_s}{dT} \right|_{T=T_{ref}} \Delta T = \frac{L_c}{R_v} \frac{e_s(T_{ref})}{T_{ref}^2} \Delta T$$

(4.43)

The linearized relationship between changes in e_s and T becomes

$$\Delta e_s \approx \gamma \left(\frac{T_{ref}^2}{\bar{T}^2} \right) \exp\left[b \left(\frac{1}{T_{ref}} - \frac{1}{\bar{T}} \right) \right] \Delta T$$

(4.44)

where $\gamma = (L_c/R_v)(e_s(T_{ref})/T_{ref}^2)(= 231 \, \mathrm{K^{-1}Pa})$ and $b = (L_c/R_v)$. After some manipulations, assuming $T_{ref}^2 \approx \bar{T}^2$, and using specific humidity $(\Delta q = \Delta e/p_s)$, the changes in latent heating are related to those in surface evaporation according to

$$L\Delta q \approx \gamma^* \exp\left(\frac{\bar{T} - T_{ref}}{\bar{T}T_{ref}/b} \right) \Delta T$$

(4.45)

where $\gamma^* = \gamma L_c/p_s$ $(\approx 5.67 \times 10^3 \, \mathrm{m^2 s^{-1} K^{-1}})$. In such a relationship, the latent heating is solely determined by the SST change regardless of the atmospheric conditions, and all changes in saturation vapor pressure associated with the SST change are converted to latent heating. Nevertheless, the slowly varying latent heating tends to follow the SST change; thus, Eq. (4.45) can be adopted for studies of low-frequency tropical phenomena.

In the ZC87 model, the vertical profile of heating in the atmosphere is assumed to vary in height as $\sin(\pi z/z_T)$, in which $z_T = 10 \, \mathrm{km}$ represents the tropopause. If the heating given by Eq. (4.45) is distributed throughout the troposphere with such a vertical profile, and the average lifetime of convective processes is τ (approximately 7 hours), then the heating at mid-troposphere per unit surface area and unit time is

$$\dot{Q}_{2(s)} = \frac{\pi}{20}\frac{1}{\tau}L\Delta q \tag{4.46}$$

Replacing the values selected for all parameters (e.g., $T_{ref} = 30°C$), the heating by convective processes can be estimated by

$$\dot{Q}_{2(s)} = \Gamma \exp\left(\frac{\bar{T} - 30°C}{17.0}\right)\Delta T \tag{4.47}$$

where $\Gamma = 0.035\,\mathrm{m^2 s^{-3} K^{-1}}$.

In addition to convective processes, low-level moisture convergence can contribute to moist the air column. In this case, the latent heat released at the mid-troposphere from the moisture convergence can be expressed as follows:

$$\dot{Q}_{2(c)} = L_c q_s M\left(-\nabla \cdot \mathbf{v}\right)(\pi/20) \tag{4.48}$$

where $M(x) = x$ if $x > 0$, otherwise $M(x) = 0$, indicating that only moisture convergence produces latent heating, and q_s is moisture in the boundary layer. Therefore, the anomalous latent heating by moisture convergence becomes

$$\dot{Q}_{2(c)} = L_c q_s [M(-\nabla \cdot \bar{\mathbf{v}} - \nabla \cdot \mathbf{v}') - M(-\nabla \cdot \bar{\mathbf{v}})](\pi/20) \tag{4.49}$$

Thus, the anomalous latent heat at the mid-troposphere in the tropical atmosphere can be written as,

$$\dot{Q}_2 = \dot{Q}_{2(s)} + \dot{Q}_{2(c)} \tag{4.50}$$

The first term in Eq. (4.50) is determined directly by the SST change, whereas the second term indicates the atmospheric response induced by the initial SST change. Therefore, in ZC87, the first guess for the latent heating is determined by $\dot{Q}_{2(s)}$, which drives atmospheric winds that can result in additional latent heating due to moisture convergence. Such a process resembles the convective instability of the second kind mechanism [Holton, 2012].

4.5.2 *Ocean component*

The single layer in the reduced gravity model formulation is not a good approximation for ocean motions near the surface. In these locations, the wind stress directly drives the ocean currents; however, the effects of wind stress strongly decrease away from the surface. In the ZC87 model, the upper ocean is considered as a two-layer system without density stratification. In this framework, the pressure gradient is the same in both layers (i.e., $\nabla P_1 = \nabla P_2$).

Therefore, the motions in both the surface and subsurface layers have one component that depends on the common pressure gradient (barotropic) while the surface layer has an additional component that is directly driven by the wind stress. The corresponding momentum equations for the surface and subsurface layers (subscripts 1 and 2, respectively) are,

$$\frac{D\mathbf{v}_1}{Dt} + f\,\mathbf{k} \times \mathbf{v}_1 = -\frac{1}{\rho}\nabla P_1 + \frac{\vec{\tau}_s}{\rho H_1} \tag{4.51a}$$

$$\frac{D\mathbf{v}_2}{Dt} + f\,\mathbf{k} \times \mathbf{v}_2 = -\frac{1}{\rho}\nabla P_1 \tag{4.51b}$$

Multiplication of Eq. (4.51a) by H_1 and of Eq. (4.51b) by H_2 and division by $H_1 + H_2 = H$ yields

$$\frac{D\mathbf{v}}{Dt} + f\,\mathbf{k} \times \mathbf{v} = -\frac{1}{\rho}\nabla P_1 + \frac{\vec{\tau}_s}{\rho H} \tag{4.52}$$

where the averaged current in the upper ocean $\mathbf{v}$ is $\mathbf{v} = (H_1\mathbf{v}_1 + H_2\mathbf{v}_2)/H$. Subtracting Eq. (4.51b) from Eq. (4.51a) gives an equation for the shear current $(\mathbf{v}_s = \mathbf{v}_1 - \mathbf{v}_2)$:

$$\frac{D\mathbf{v}_s}{Dt} + f\,\mathbf{k} \times \mathbf{v}_s = \frac{\vec{\tau}_s}{\rho H_1} \tag{4.53}$$

The shear current is solely determined by the surface wind stress. It can be assumed that the adjustment of the shear current to the wind is fast enough to allow for neglecting the tendency term. Accordingly, Eq. (4.53) becomes

$$r_s\mathbf{v}_s + f\,\mathbf{k} \times \mathbf{v}_s = \frac{\vec{\tau}_s}{\rho H_1} \tag{4.54}$$

which is identical to the expression satisfied by an Ekman current. The velocity in the surface layer is then given by $\mathbf{v}_1 = \mathbf{v} + H_1\mathbf{v}_s/H$.

The temperature equation for the surface layer (T or SST) is

$$\frac{\partial T}{\partial t} = \frac{Q_{net}}{\rho C_p H_1} - \mathbf{v}_1 \cdot \nabla T - M(w_s)\frac{(T - T_{sub})}{H_1} \tag{4.55}$$

where $M(x) = x$ when $x > 0$ and 0 otherwise. The net surface heat flux (Q_{net}) include contributions from net solar and longwave radiative fluxes, as well as latent and sensible turbulent heat fluxes. The radiative flux depends on air temperature, humidity, and cloudiness; the turbulent heat flux is controlled by the wind speed, and temperature and humidity differences at the air-sea interface. In the tropical ocean, Q_{net} is almost linearly proportional to the surface temperature. Thus, the expression for the net heat flux over the tropical

Pacific could be approximated with a Newtonian cooling parameterization: $Q_{net}/\rho C_P H_s \approx -\alpha_s T$, where ρ and C_P are density and specific heat of seawater, respectively. The upwelling velocity is given by the convergence of the surface layer current, $w_s = H_1 \nabla \cdot \mathbf{v}_1$, and T_{sub} is the temperature in the subsurface. In this formulation, the vertical advection effect is taken into account in locations of upwelling

ZC87 is intended to simulate the behavior of SST anomalies about climatological mean conditions prescribed according to the observation. Thus, Eq. (4.55) is linearized with respect to the climatological mean value. If primes indicate anomalies about the climatological mean, we can write,

$$\frac{\partial T'}{\partial t} = -\mathbf{v'}_1 \cdot \nabla(\bar{T} + T') - \bar{\mathbf{v}}_1 \cdot \nabla T' - [M(\bar{w}_s + w'_s) - M(\bar{w}_s)]\bar{T}_z$$

$$- \gamma_e M(\bar{w}_s)(T' - T'_{sub})/H_S - \alpha_s T' \tag{4.56}$$

where γ_e is a mixing efficiency coefficient (0.75) and $\bar{T}_z$ is the prescribed mean vertical temperature gradient. H_s is set to $50\,\text{m}$ over the entire domain. This is a reasonable approximation for the eastern Pacific. Over the western Pacific, however, the surface layer is usually deeper than $50\,\text{m}$. The underestimation of the difference between SST and subsurface temperature in the region may lead to an underestimation of the vertical thermal advection by upwelling. This error may be negligible over the western Pacific, however, where the local thermocline is deep and vertical thermal advection is also weak.

The anomalous subsurface temperature T'_{sub} is empirically parameterized as a function of the mean and anomalous thermocline depth. This functional relation is based on approximating the functional dependence of temperature with depth over the tropical Pacific by means of the hyperbolic tangent function [e.g., Zebiak and Cane, 1987; Battisti, 1988]:

$$T'_{sub} = A(\tanh[B(\bar{h} + h')] - \tanh[B\bar{h}]) \tag{4.57}$$

where $A = 29°\text{C}$ and $B = (80\,\text{m})^{-1}$ for $h' > 0$, and $A = -40°\text{C}$ and $B = (33\,\text{m})^{-1}$ for $h' < 0$. These parameters have been recalculated using Expendable Bathy Thermograph (XBT) observations by Dewitte and Perigaud [1996], in which the fitted parameters vary with longitude. The modified parameters for $h' > 0$ are as follows: $A = 30°\text{C}$ west of $130°\text{W}$, and $A = 40°\text{C}$ east of $100°\text{W}$; $B = (110\,\text{m})^{-1}$ west of $130°\text{W}$ and $B = (30\,\text{m})^{-1}$ east of $100°\text{W}$, where A and B increase linearly between $130°$ and $100°\text{W}$. In turn, for $h' < 0$, A has the opposite sign while B does not change signs. In this case, the variations in T'_{sub} are symmetric with respect to h', and their sensitivity increases to the east.

Finally, in ZC87, the equation for the surface wind stress anomalies acting on the ocean is

$$\vec{\tau}' = \rho_a C_D |\bar{\mathbf{v}} + \mathbf{v}'|(\bar{\mathbf{v}} + \mathbf{v}') - \rho_a C_D |\bar{\mathbf{v}}|\bar{\mathbf{v}} \tag{4.58}$$

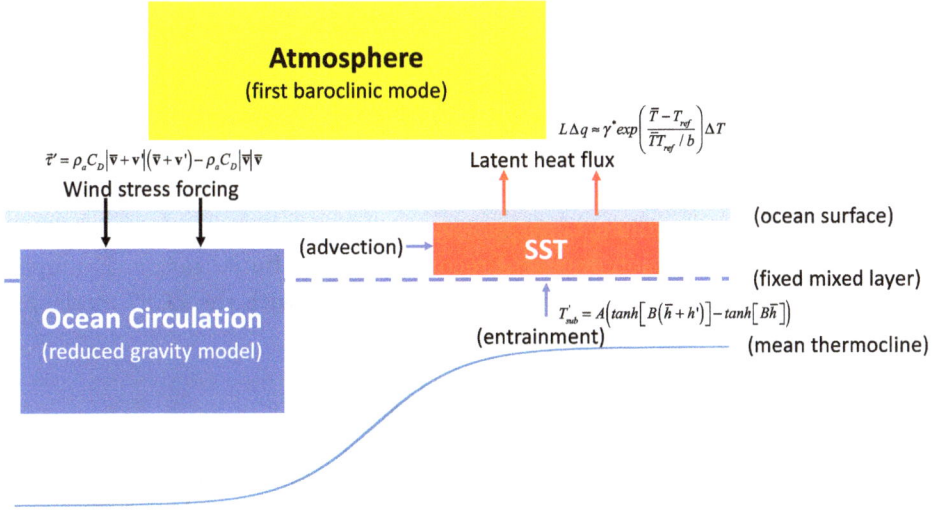

Figure 4.4.   Schematic diagram of ZC87 model structure, and atmosphere-ocean coupling process.

where $\bar{\mathbf{v}}$ and ρ_a are the monthly climatological wind vector and air density at the lowest model level, respectively, and C_D is the drag coefficient (see Ch. 5). Zebiak [1984] takes $\rho_a C_D = 3.2 \times 10^{-3} \, \mathrm{kg \, m^{-3}}$.

4.5.3   *Atmosphere-ocean coupling*

In the ZC87 model the anomalous wind stress (Eq. (4.58)) drives the three-dimensional upper-ocean circulation including thermocline depth anomalies (Eq. (4.39c) and surface currents (Eqs. (4.39) and (4.54)), and thermal advection by these oceanic circulations mainly control the SST perturbations (Eq. (4.56)). Meanwhile, the SST perturbations induce the atmospheric circulation by providing latent heating as atmospheric forcing (Eq. (4.50)), which drives the atmospheric circulation (Eq. (4.6)). In this way, the atmosphere and ocean are tightly coupled (see Fig. 4.4).

4.5.4   *ENSO predictions with the ZC87 model*

The first attempt at dynamical forecasting of ENSO using the ZC87 model was quite successful, even though the model was much simpler than several operation weather prediction systems at the time. The results of that exercise suggested that ENSO could be predictable one or two years in advance [Cane et al., 1986]. Although other models have been applied to ENSO predictions, the ZC87 model is still being used for ENSO forecasting, even in operational contexts (e.g., (https://iri.columbia.edu/our-expertise/climate/forecasts/enso/current/?enso_tab=enso-sst_table).

The ZC87 model has been modified several times since it was first applied to forecasting and data assimilation procedures have steadily improved. Inspite of this, the model's forecast skill was lower during the 1990s than during the 1980s (see Fig. 4.6). In particular, the model failed to predict the 1997–98 El Niño [Chen et al., 1998]. This motivated increased attention on two key components of a prediction system: Preconditions and Initialization. Examples of the importance of these components in predictions with the ZC87 model are given in the following section.

4.6   Preconditions and initialization

In long-term forecasting with a numerical model, incomplete physics is the primary contributor to limiting the forecast skill; in short-term forecasting skill is most likely limited by difficulties with the initialization procedure. Errors in the initial conditions may be reduced by assimilating the observed data into the coupled model. For example, a stand-alone assimilation, such as assimilating observed ocean data into the ocean component without coupling to the atmosphere component, may reduce the errors in the initial condition for the ocean; however, when the ocean and atmosphere are coupled to start the forecast, an initial shock may occur and deteriorate the model's performance. In this section, we just touch upon two important aspects of the challenging issues in prediction: (1) the preconditions for El Niño development, and (2) the importance of model initialization.

4.6.1   *Preconditions*

In either average or La Niña years, the easterly trade winds in the equatorial Pacific push the surface warm water to the west leading to a westward slanted slope of sea level along the equator. Figures 4.5(a) and 4.5(b) illustrate that the accumulation of warm water over the tropical western Pacific tends to be larger during boreal spring of years in which the warm ENSO phase develops [Wyrtki 1975]. In addition, Figure 4.5(a) shows that during springs a close relationship exists between higher heat content and surface westerlies over the tropical western Pacific.

In view of these associations, the accumulated heat content in the upper ocean of the tropical western Pacific [Wyrtki 1975; Jin 1997; see Sec. 3.3 in this book] and westerly wind bursts (WWBs) over the western-to-central Pacific [Harrison and Vecchi, 1997; McPhaden and Yu, 1999] during boreal spring have been proposed as proper preconditions for El Niño development. WWBs are episodic events of westerly surface winds with a strength of 5 to $7\,\mathrm{ms}^{-1}$,

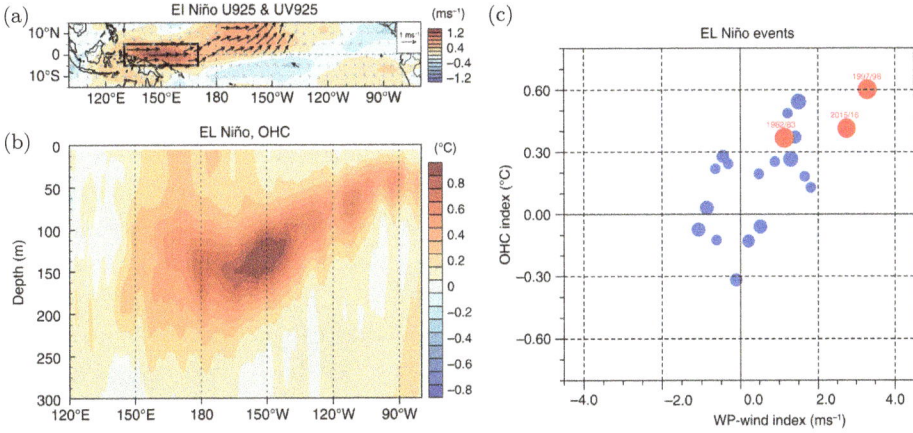

Figure 4.5.   Composite maps of the 925 hPa anomalous (a) zonal wind (shadings, at intervals of 0.1 m/s) and horizontal winds (vectors, wind speed >0.5 m/s highlighted by thickened black vectors) and (b) upper ocean temperature at intervals of 0.1°C averaged over the equatorial band of 5°S–5°N during Feb–Mar–Apr(0) for the 20 El Niño events for the period of 1958–2016. (c) 925 hPa zonal wind anomalies (WP-wind) (130°–170°E, 5°S–5°N; x-axis) and upper 300 m depth average temperature anomalies (ocean heat content, OHC) (120°E–80°W, 5°S–5°N; y-axis) indices during Feb–Mar–Apr (0) for the respective El Niño events. Scatters indicate each of the 20 El Niño events, and the size of markers is proportional to the El Niño amplitudes as following the Niño-3.4 index intensities during Dec–Jan–Feb (0/1). Extreme (moderate) El Niño events are marked with red (blue) markers. (Figure 1 in Kim and An [2018].)

zonal extent of 20°–40°, duration of 5–30 days, and frequency of approximately 5 to 10 times per year [An et al., 2020]. Heat accumulation in the western Pacific acts as energy reservoir for El Niño while WWBs play the role of communicator to deliver the accumulated energy from the western to the eastern Pacific and that of further booster via state-dependent noise [Eisenman et al., 2005]. In particular, strong events (i.e., 1982–83, 1997–98, and 2015–16) required larger heat content and stronger westerly anomalies over the western Pacific as precursors (Fig. 4.5(c)).

However, El Niño has occurred even under very low heat content and weak westerly anomalies; thus, oceanic heat accumulation is a necessary, but not a sufficient condition. Moreover, the linear relationship between the accumulated heat amount and El Niño intensity frequently breaks down, for example in the 2014–15 El Niño. Heat accumulation during the 2014 spring was comparable with that during the 1997 spring and yet the El Niño was weak during the following winter. Possible reasons for this break in the relationship have been proposed, such as the lack of WWBs or even the occurrence of easterly wind events. It was argued that strong easterly winds that developed in the following boreal summer damped the El Niño condition by generating upwelling Kelvin waves [Hu and Fedorov, 2017]. Also, weaker warming during boreal summer 2014 provided drier conditions over the central Pacific, causing a saltier ocean surface. Such positive sea surface salinity anomalies in the central Pacific prevented the warm pool from expanding westward, as it occurs in a typical El Niño event [Chi et al., 2019].

As previously mentioned, westerly wind anomalies over the equatorial western Pacific can play a key role in the outbreak of El Niño events [Yu and Rienecker, 1998; Lengaigne et al., 2004]. WWBs over that region generate a downwelling Kelvin wave that deepens the thermocline in the eastern Pacific, leading to SST warming through the reduction of upwelling of the subsurface cold water. Finally, the increased SST anomalies can further develop into an El Niño event through the Bjerknes positive feedback [Bjerknes, 1969; see Ch. 1]. However, a single WWB event may not suffice to switch on the Bjerknes feedback and a series of WWBs may be required for triggering a large-scale air-sea coupled positive feedback. WWBs were initially considered as additive stochastic forcing [Kleeman and Moore, 1997]; a strong dependency of these events on the state of the SST, particularly during a developing El Niño, was later widely reported [Verbickas, 1998; Yu et al., 2003; Eisenman et al., 2005], indicating that some of the WWBs should be treated as deterministic dynamics [An et al., 2020; also see Sec. 3.7 of this book].

Taking into account information provided by the preconditions described in the previous paragraphs can be key to the success of an ENSO forecast. As stated in Sec. 4.5.4, the ZC87 model failed to predict the 1997–98 El Niño [Chen et al., 1998]. According to Fig. 4.5(c), for example, the oceanic heat content during the 1997 spring was massive. To incorporate such information into the forecasting system Chen et al. [1998] suggested to assimilate the sea level anomalies from the tide gauge data. Specifically, the ocean height field h is modified as $h = \alpha h_a + (1 - \alpha)h_c$, where h_a and h_c are the sea-level assimilated ocean model run and the coupled model run, respectively. The nudging parameter α has a Gaussian distribution in latitude with a maximum value of 0.8 at the equator and an e-folding scale of $2°$ latitude. Thus, the nudging was done effectively only within the equatorial waveguide, where the SST anomalies were most sensitive to change in the oceanic dynamic fields. This oceanic subsurface data assimilation improved the ENSO forecast skill by correcting the model ocean state and precondition of ENSO [Chen et al., 1998].

4.6.2 *Initialization*

In earlier forecasts with the ZC87 model, only wind information was assimilated. That is, the ocean model forced with the observed, monthly-mean wind stress anomaly fields was integrated for a period up to the forecast initial time; then, the computed SST anomalies were used to run the atmospheric model

[Cane et al., 1986]. Finally, the calculated anomalies in thermocline depth, currents, SST, and surface winds were used as initial conditions for a forecast. In such a way, the coupled system was closer to balance initially and a shock could be minimized when the forecast begins.

Chen et al. [1995] introduced a coupled approach, in which both the SST and wind data were assimilated into the coupled system for forecast initialization. In this approach, the wind stress anomalies produced by the model τ_m at each time step were replaced by $\alpha\tau_o + (1 - \alpha)\tau_m$, where τ_o is the observed wind stress anomaly and α is a function of latitude. The values of α are determined according to the model's skill to simulate the winds stress anomalies, compared with the observation. This modified assimilation procedure yielded improved forecasts of El Niño, especially for the 1980s compared with previous forecasting procedure. It has been shown, however, that ENSO's predictability varies on decadal or longer time scales [see Fig. 4.6; Chen et al., 1995].

4.7 Another example of an intermediate-complexity model for El Niño prediction

Here, we examine the coupled model for El Niño predictions developed by Barnett et al. [1993]. In this case, the ocean was a nonlinear primitive equation model that computes the three-dimensional current field, sea level, and sea surface temperature on an equatorial β-plane covering the tropical Pacific (30S~30N, 130E~70W). This ocean model was coupled to a statistical atmospheric model. That is, the ocean model is more complex than those that were presented so far in this chapter, although it is not a full OGCM.

The first step in building the statistical model of the atmosphere was to decompose monthly anomaly fields of SST $(T(x, y, t))$ and surface wind stress $(\tau(x, y, t))$ from an observational dataset in their spatial and temporal components. This was done by using the empirical orthogonal functions (EOFs) method with respect to each calendar month. For a specified month,

$$T(x, y, t) = \sum_n \alpha_n(t)e_n(x, y) \tag{4.59a}$$

$$\tau(x, y, t) = \sum_n \beta_n(t)f_n(x, y) \tag{4.59b}$$

In general, the number of modes retained for estimation depends on how many modes of the surface-wind stress are directly driven by an SST anomaly. Barnett et al. [1993] suggested that three to six modes suffice for the expansion. The model equations provide the matrix of regression coefficients (C_{mn}) that

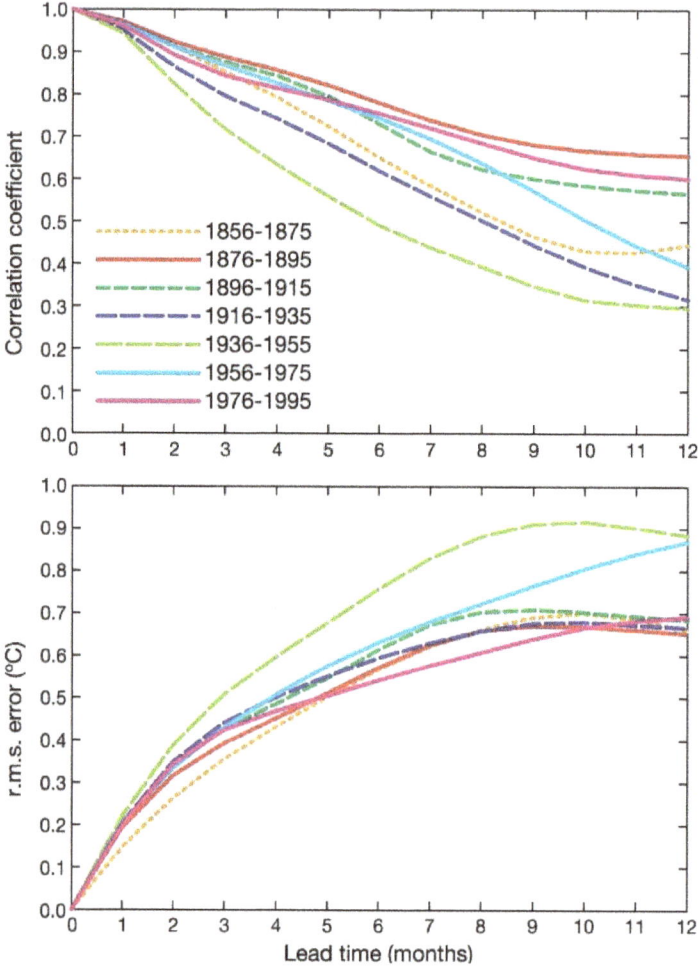

Figure 4.6.   Anomaly correlations and r.m.s. errors between the observed and the predicted values of the NINO3.4 index. The prediction has been done using LDEO (Lamont-Doherty Earth Observatory) ENSO forecast model, which is based on ZC87). (Figure 2 in Chen et al. [2004].)

connects the two variables,

$$C_{nm} = \frac{\langle \alpha_m \beta_n \rangle}{\langle \alpha_m^2 \rangle} \tag{4.60}$$

where angle brackets denote time average.

In running the coupled model, the simulated SST anomaly $\tilde{T}(x, y, t)$ is projected onto the base functions $e_n(x, y)$ to compute

$$\tilde{\alpha}_n(t) = \sum_{x,y} \tilde{T}(x, y, t) e_n(x, y) \tag{4.61a}$$

which through C_{nm} gives,

$$\hat{\beta}_m(t) = \sum_n C_{nm}\tilde{\alpha}_n(t) \tag{4.61b}$$

and from there the wind stress anomaly can be obtained,

$$\hat{\tau}(x,y,t) = \sum_m \hat{\beta}_m(t) f_m(x,y) \tag{4.62}$$

The model described successfully reproduced many observed features of El Niño. Hindcast and forecast experiments by the model demonstrated useful skill at a lead time of up to 18 months, especially over the central equatorial Pacific.

Most coupled models having a linear statistical model for the atmosphere [Barnett et al., 1993; Syu and Neelin, 2000; Balmaseda et al., 1994] are unable to capture nonlinear aspects of atmosphere-ocean interactions. Tang and Hsieh [2002] developed a dynamical ocean model coupled to a nonlinear neural network atmosphere and compared to that coupled to a linear regression model. They found that the model with the nonlinear neural network atmosphere is more successful in capturing the nonlinear nature in air-sea coupled system than a linear regression model. Especially the ENSO predictability in the 1990s by a nonlinear neural network model did not decrease as it does in a linear regression model.

4.8 Perspectives

Intensive theoretical and modeling efforts supported by remarkable improvements in the capabilities to gather and analyze observational data in the last four decades have all contributed to a better understanding of the fundamental mechanisms of ENSO. Along with these accomplishments, significant progress in ENSO prediction has been achieved [e.g., Latif et al., 1994; Jin et al., 2008]. Predictions of ENSO warm and cold events with different types of numerical models have proved useful for lead times of up to 6–12 months [Jin et al., 2008].

The forecast skill of these prediction models is not uniform/consistent but may depend on ENSO phase and intensity as well as on peculiar features of the event. In general, a stronger event is better predicted than a weaker one because of the stronger signal to noise ratio. Initiation of an event is generally difficult to capture, but once this is accomplished then the ensuing evolution is reasonably well predicted. Nevertheless, the skill in predicting the amplitude on an event and its decaying phase are still low at the present time. The forecast skill usually drops in boreal spring, a feature that is often referred to as the "spring predictability barrier". During boreal spring, the coupling between atmosphere and ocean becomes weaker and random atmospheric processes are a stronger influence on SST than ocean dynamic processes. Nevertheless, it is unclear whether the spring barrier reflects an intrinsic property of the real climate system or whether it is exacerbated by the deficiencies of the forecast models [e.g., Chen et al., 1995]. Recently, Chen et al. [2020] introduced a method to noticeably increase the ENSO prediction skill beyond the spring predictability barrier and proposed that this feature is related to tropical–extratropical interactions. During the last two decades, the extratropical impact of climate variability in the tropics has intensified, which has led to a reduction of ENSO predictability. Therefore, by considering a longer lead-time of the extratropical–tropical ocean-to-atmosphere interaction process, the prediction skill is clearly improved beyond the spring barrier.

Beginning in early 2002, forecasts of the Niño-3.4 index (defined as the SST anomalies averaged over 5°N–5°S, 120°–170°W) from various models have been collected and displayed monthly at an International Research Institute for Climate and Society web page (http://iri.columbia.edu/climate/ENSO/curren tinfo/SST_table.html). A couple of decades ago, forecast with different types of models had roughly the same level of skill [Kirtman et al., 2001]. At present, more than 20 real-time forecasting models are participating in this ENSO forecasting collection program. Seasonal forecasts with comprehensive CGCMs (see Ch. 7) have been continuously improving, especially if they are performed in ensemble mode and sophisticated data assimilation schemes are used [e.g., Mason et al., 1999; Kanamitsu et al., 2002]. Currently, the forecast skill of dynamical models exceeds that of statistical models [Baroston et al., 2012].

Despite all efforts made, answering questions such as "whether El Niño will occur" and "if so, how significant will it be" remain challenging issues in ENSO prediction [L'Heureux et al., 2017]. There is still ample room for improvement in forecast skill of intermediate coupled models. Additional upgrades in model initialization and data assimilation, the better simulation of surface heat and freshwater fluxes, and improved representation of relevant processes outside of the tropical Pacific could all lead to improved ENSO forecasts [Chen and Cane, 2008].

Chapter 5

AGCMs Coupled to Simpler Ocean Models

5.1 Introduction

This chapter is dedicated to a family of coupled atmosphere-ocean systems in which the atmospheric component is a general circulation model (AGCM) and the ocean component is a simpler model. The simpler ocean models have already been mentioned in previous chapters. AGCMs synthesize a number of mutually interacting processes in the atmosphere that are deemed as essential for climate. Some outstanding features of AGCMs are briefly reviewed in this chapter. Particular attention is dedicated to the determination of fields to be passed to the ocean model in order to realize the coupling.

5.2 AGCMs

AGCMs are based on the equations of fluid motion, conservation of water and other substances, the laws of thermodynamics, and detailed physics of radiation transfer in the atmosphere. The models include realistic representations of geography and surface conditions. They are forced at the top by solar radiation generally given as time varying functions of latitude and longitude.

In building an AGCM, the partial differential equations that govern fluid motions are replaced by a finite number of algebraic equations. Further, continuous functions are represented by their values at a discrete number of points. This brings up the discretization problem, which may result in serious distortions of the continuous system solutions at any grid resolution. It is impossible for any model of the atmosphere to resolve the entire spectrum of atmospheric phenomena ranging from millimeter (viscous subrange) to planetary scales ($\sim 10^4$ km). The collective effects of unresolved processes, therefore, must be formulated in terms of the resolvable-scale predicted variables. This is the problem of parameterization, which represents a crucial part of general circulation modeling. In this chapter, our emphasis is on the physical processes that are directly relevant to the calculation of fluxes at the atmosphere-ocean interface.

In the vertical, most AGCMs use the σ coordinate, where σ is pressure normalized by surface pressure, or a variant such as a hybrid coordinate with σ

changing to pressure above a certain level. This choice of coordinates simplifies
the model's lower boundary atmosphere as a coordinate surface even above
irregular terrain, and the kinematic lower boundary condition in an AGCM is a
statement that Earth's surface is a material surface. This simplification comes
at the possible cost of introducing discretization errors. An error of particular
concern can occur in the horizontal pressure gradient force in the momentum
equation at locations where the topography is especially steep.

In Ch. 4 we introduced a two-level model of the atmosphere in which
horizontal velocities were defined in each level, potential temperature as well as
vertical velocities were defined at mid-level, and vertical velocities were defined
at the top and bottom boundaries (see Fig. 4.1). The Mintz–Arakawa model of
the global atmosphere, one of the first AGCMs built in the early 1960s, was also
a two-level model with a different distribution of variables in the vertical level
than the one in Fig. 4.1 [see, e.g., Johnson and Arakawa, 1996]. In this model,
horizontal velocities and potential temperature were both defined at each level.
The top level was set at 200 hPa, and as such it was labeled as a "tropospheric
model". The seasonal changes in solar radiation and long-wave cooling for each
level were prescribed on the basis of observational data and consequently some
effects of clouds were included although these were not present as moisture was
not predicted. In this way, the Mintz–Arawaka model could generate its own
"climate", and could be appropriately labeled as a "general circulation model".

Figure 5.1 shows the vertical distribution of prognostic variables in most
AGCMs. Models that use the Lorenz grid shown in Fig. 5.1(a) — such as Mintz–
Arakawa — define horizontal velocity $\mathbf{v}$ and potential temperature θ at the same

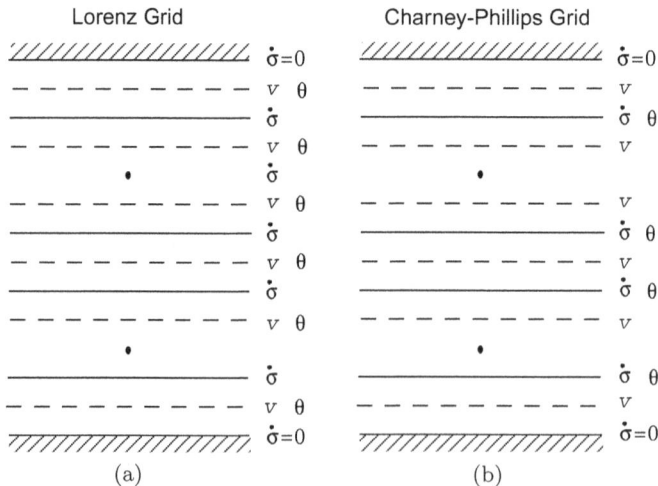

Figure 5.1.   Vertical distribution of prognostic variables in most AGCMs. Symbols are
defined in the text. (From Fig. 5 in Mechoso and Arakawa [2015].)

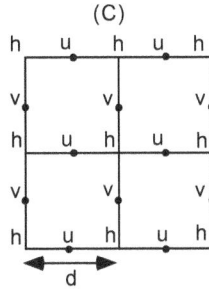

Figure 5.2.   The Arakawa C-grid used in many contemporary AGCMs. (From Figure 4 in Mechoso and Arakawa [2015].)

level, while "vertical velocity" $\dot{\sigma} = D\sigma/Dt$ is defined at mid-levels. Models that use the Charney–Phillips grid shown in Fig. 5.1(b) — predict θ at mid-levels. The merits and demerits of the Lorenz and Charney grids are discussed in Arakawa and Konor [1996]. Contemporary AGCMs can use scores of vertical levels and the limitation on this number is given primarily by the validity of the approximations used to derive the prognostic equations, as well as by the computer power available to perform the calculations.

The governing equations are also discretized on surfaces where the vertical coordinate is constant. In this chapter we are primarily concerned with the horizontal grids in which physical processes (e.g., convection) are discretized. In a class of AGCMs both dynamical (e.g., advection) and physical processes are computed on longitude-latitude grids based either on the spherical coordinates or on quasi-uniform spherical "geodesic" grids generated from icosahedra or other platonic solids. Another class of AGCMs use the spectral method to solve dynamical processes other than advection, but use a grid for the physical processes. Figure 5.2 shows the distribution of variables in the Arakawa C-grid [Arakawa and Lamb, 1977] used in many AGCMs. In this figure, the geopotential height h is a synonym for pressure. Note that the horizontal velocity components are not defined at the same location. This means that some spatial averaging is required for the calculation of the crucial Coriolis force term in the momentum equations. Such an averaging is acceptable for atmospheric motions, of which the key ones are in scales significantly longer than the grid.

5.3   Fluxes at the atmosphere-ocean interface

When coupled to an ocean model, the AGCM provides to the ocean model the surface flux of momentum (stress, $\mathbf{F_v}$), heat (F_θ), water (F_q), and other material properties. In Fig. 5.3, these fluxes have been combined under the label F_{turb} because they are determined by turbulent processes. The AGCM also provides to the ocean fluxes due to radiation energy from several sources,

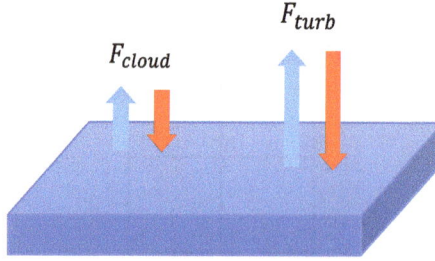

Figure 5.3. Conceptual representation of fluxes at the atmosphere-ocean interface. (Figure created by author.)

including the sun and gases that form the atmosphere. These fluxes depend on the clouds overhead and hence they have been labeled as F_{cloud}. AGCMs may also consider the aerosol fluxes at the surface. Although these are very important particularly for climate change studies, we will not discuss them in detail in the present book.

The standard practice in AGCMs is to calculate the turbulent surface fluxes based on the bulk aerodynamical method using the resulting variables at the surface and at a level L above the surface (to be defined). In this approach, the turbulent fluxes are given by

$$(\mathbf{F_v})_S \equiv -\rho_S C_D |\mathbf{v}_L| \mathbf{v}_L$$

$$(F_\theta)_S \equiv -\rho_S C_H C_p |\mathbf{v}_L| (\theta_L - \theta_S) \tag{5.1}$$

$$(F_q)_S \equiv -\rho_S C_q |\mathbf{v}_L| k[q_L - q^*(T_S, p_S)]$$

where $\mathbf{v}_L, \theta_L, q_L$ are horizontal velocity, potential temperature, and mixing ratio of water vapor, respectively, at level L to be defined in each case; ρ_S, T_S, p_S are atmospheric density, surface temperature, and surface pressure, respectively, C_p is specific heat of air, k represents ground wetness ($k = 1$ over the ocean), and q^* is saturation mixing ratio. The exchange coefficients C_D, C_H, C_q are computed in a highly specialized model component known as the planetary boundary layer (PBL) parameterization.

In the following sections we provide a rationale for using Eq. (5.1) in AGCMs. We will examine methods to estimate the values of the coefficients C_D, C_H, C_q, which we can anticipate as resulting from a combination of theory and experimentation in the laboratory or in the field. The presentation will clarify what is meant by the "L" level.

5.4   Calculation of fluxes at the atmosphere-ocean interface

We start by introducing a framework for the flow of a stratified layer of the atmosphere along the surface. In this layer, turbulence processes play key roles in the transports and budgets of momentum, heat and moisture. Turbulence is characterized by irregular fluctuations in space and time of fluid properties.

Let us consider a flow variable $A(P,t)$, where $P = P(x,y,z)$ and t is time. We define the mean in time, $\bar{A}(P,t)$, in the following way,

$$\overline{A}(P,t) = \frac{1}{T} \int_{t-T/2}^{t+T/2} A(P,\tau)d\tau \tag{5.2}$$

where T is a "time-window" that represents the time scale of the "slow-changing" motions. Thus, for instantaneous values,

$$A(P,t) = \overline{A}(P,t) + A'(P,t) \tag{5.3}$$

The primed quantity in Eq. (5.3) represents the "eddy motions". It is generally assumed that the scales of the mean and eddy motions are well separated and that T is chosen such that $\overline{(\overline{A})} = \bar{A}$, therefore $\overline{A'} = 0$. This is referred to as the *Reynold's averaging*. Moreover, if $A(P,t)$ and $B(P,t)$ are both flow variables, it is easy to show that $\overline{A\,B} = \overline{A}\,\overline{B} + \overline{A'\,B'}$. Note that $\overline{A'\,B'}$ is the covariance of A and B.

Using the properties of the Reynolds averaging, we can write the Reynolds equation for turbulent motions in the following way,

$$\frac{D\bar{V}}{Dt} = -\frac{1}{\rho_0}\nabla \cdot T_{\mathrm{R}} - 2\Omega \times \bar{V} - \frac{\nabla \bar{p}}{\rho_0} + e_3 \frac{\bar{\theta}}{\theta_0} g \tag{5.4}$$

$$\frac{D}{Dt} = \frac{\partial}{\partial t} + \bar{V} \cdot \nabla \tag{5.5}$$

where V is the (three-dimensional) velocity, p is pressure, θ is potential temperature, Ω is rotation rate, g is gravity, t is time; and e_i is the unit vector in the direction $x(i = 1), y(i = 2)$ and $z(i = 3)$. The anelastic approximation has been used and the quantities with subscript "0" are reference values. T_{R} in Eq. (5.4) is the Reynolds stress tensor, which represents the effects of the turbulent motions on the mean motion. This tensor is given by

$$(T_{\mathrm{R}})_{i.j} = \rho_0 \overline{v'_i v'_j}$$

$$T_{\mathrm{R}} \equiv - \begin{bmatrix} \overline{u'^2} & \overline{u'v'} & \overline{u'w'} \\ \overline{u'v'} & \overline{v'^2} & \overline{v'w'} \\ \overline{u'w'} & \overline{v'w'} & \overline{w'^2} \end{bmatrix} \tag{5.6}$$

$$\nabla \cdot T_{\mathrm{R}} = \sum_i \frac{\partial}{\partial x_i}(T_{\mathrm{R}} \cdot e_i)\, e_i$$

where $V' = u'e_1 + v'e_2 + w'e_3$. Similarly, for the potential temperature, we have

$$\frac{D\bar{\theta}}{Dt} = -\frac{1}{\rho_0}(\rho_0 \overline{V'\theta'}) + \frac{1}{c_p}\frac{\theta_0}{T_0}\bar{Q} \tag{5.7}$$

where $\bar{Q}$ is diabatic heating rate. The eddy kinetic energy equation is

$$\frac{D}{Dt}\frac{1}{2}\overline{V'^2} = -\frac{1}{\rho_0}\nabla\cdot\left(\overline{\rho_0 V'\frac{1}{2}V'^2 + \overline{V'p'}}\right) - \sum_{i,j}\overline{v_i'v_j'}\frac{\partial\bar{v}_i}{\partial x_j} + \frac{g}{\theta_0}\overline{v_3'\theta'} - \varepsilon_M \quad (5.8)$$

where the last term in the right-hand side is the dissipation of eddy kinetic energy. For the thermodynamic equation we have

$$\frac{D}{Dt}\frac{1}{2}\overline{\theta'^2} = -\frac{1}{\rho_0}\nabla\cdot\left(\overline{\rho_0 V'\frac{1}{2}\theta'^2}\right) - \overline{V'\theta'}\cdot\nabla\bar{\theta} + \frac{1}{c_p}\frac{\theta_0}{T_0}\overline{\theta'Q'} - \varepsilon_\theta \quad (5.9)$$

In the applications addressed in this chapter, it is usual to assume that only the vertical transports by turbulence are important. Moreover, hydrostatic balance is assumed according to which the vertical velocity component is no longer a prognostic variable. The kinetic energy and thermodynamic equations become

$$\frac{D}{Dt}\frac{1}{2}\overline{\mathbf{v}'^2} = \frac{1}{\rho_0}\frac{\partial}{\partial z}\left(\overline{\rho_0 w'\frac{1}{2}\mathbf{v}'^2 + \overline{w'p'}}\right) - \overline{w'\mathbf{v}'}\cdot\frac{\partial\bar{\mathbf{v}}}{\partial z} + \frac{g}{\theta_0}\overline{w'\theta'} - \varepsilon_M \quad (5.10)$$

$$\frac{D}{Dt}\frac{1}{2}\overline{\theta'^2} = -\frac{1}{\rho_0}\frac{\partial}{\partial z}\left(\overline{\rho_0 w'\frac{1}{2}\theta'^2}\right) - \overline{w'\theta'}\cdot\nabla\bar{\theta} + \frac{1}{c_v}\frac{\theta_0}{T_0}\overline{\theta'Q'} - \varepsilon_\theta \quad (5.11)$$

In Eqs. (5.10) and (5.11) the D/Dt operator is defined as

$$\frac{D}{Dt} = \frac{\partial}{\partial t} + \mathbf{v}\cdot\nabla \quad (5.12)$$

In the eddy kinetic energy equation, the second and third terms in the right-hand side represent the gain or loss of kinetic energy through shear and buoyancy processes, respectively. These are very important terms in the analysis that follows. We will distinguish the following three cases, according to whether the buoyancy force contributes to gain or lose eddy kinetic energy:

(1) $\overline{w'\theta'} = 0$, or *neutral case*
(2) $\overline{w'\theta'} < 0$, or *stable case*
(3) $\overline{w'\theta'} > 0$, or *unstable case*

5.5   The surface layer in the atmosphere

Moving away from the material surface at the bottom of the atmosphere one first finds a viscous layer. Above this comes the *surface layer*, where the vertical turbulent fluxes of horizontal velocity (and horizontal momentum), potential temperature and moisture can be considered constant in the vertical direction. The surface layer is thin, usually less than $100\,\mathrm{m}$ deep. Numerical models must have very high resolution to resolve such a layer, in which turbulent eddies are in the inertial subrange and gain their energy from large eddies, not directly from the mean flow, and viscosity can be neglected. The challenge is particularly severe for numerical climate models, which cover a global domain. For the purposes of these models, the interface between atmosphere and material surfaces — i.e., the ocean, sea-ice, or ground — is represented by the surface layer that receives information from the viscous sublayer on boundary conditions (e.g., non-slip) and from which the outer boundary layer removes turbulent fluxes. To address these issues, climate models do not attempt to resolve the surface layer but use closure assumptions to evaluate the vertical turbulent fluxes.

In the surface layer, vertical fluxes of a flow variable $A(P, t)$ (e.g., horizontal velocity, potential temperature, and moisture) are assumed to be constant in height. To see this, we consider the following prognostic equation for the time variations of $\bar{A}$,

$$\rho_0 \left(\frac{\partial \bar{A}}{\partial t} + \ldots\ldots \right) = -\frac{\partial}{\partial z}(\rho_0 \overline{w' A'}) \tag{5.13}$$

Integration of Eq. (5.13) from $z = z_0$ to $z = z_0 + \Delta z$ yields

$$\rho_0 \Delta z \left(\frac{\partial \bar{A}}{\partial t} + \ldots\ldots \right) = -\left[(\rho_0 \overline{w' A'})_{z_0 + \Delta z} - (\rho_0 \overline{w' A'})_{z_0} \right]$$

Therefore, for small values of Δz,

$$(\overline{\rho_0 w' A'})_{z_0 + \Delta z} \cong (\overline{\rho_0 w' A'})_{z_0} \tag{5.14}$$

5.5.1   *The Monin–Obukhov theory and fluxes at the surface*

In the stratified surface layer we can determine, through the Monin–Obukhov theory, the existence of universal functions that allow for the computation of the turbulent fluxes from information about the structure of the layer in reference to flow conditions at a certain height. The theory is formally based on a non-dimensional analysis. To simplify the notation, we will take a flow in which the mean velocity is only in the x-direction.

The basic assumption in the Monin–Obukhov theory is that the turbulent flow in the surface layer is governed by equations of the form,

$$H_m(\overline{w' u'}, \overline{w' \theta'}, \partial \bar{u}/\partial z, z) = 0$$
$$H_h(\overline{w' u'}, \overline{w' \theta'}, \partial \bar{\theta}/\partial z, z) = 0 \tag{5.15}$$

Building non-dimensional expressions with the arguments of H_m, H_h, we can write

$$\frac{\kappa \zeta}{u_*} \frac{\partial \bar{u}}{\partial \zeta} = \phi_m(\zeta) \quad \frac{\kappa \zeta}{\theta_*} \frac{\partial \bar{\theta}}{\partial \zeta} = \phi_h(\zeta) \tag{5.16}$$

In Eq. (5.16) κ is the von Karman constant (see Appendix), and

$$u_*^2 \equiv -\overline{w'u'} \quad u_*\theta_* \equiv -\overline{w'\theta'} \quad \zeta \equiv z/L \tag{5.17}$$

where L is the Monin–Obukhov length, which is defined by

$$L \equiv -\frac{u_*^3}{\kappa \frac{g}{\theta_0}\overline{w'\theta'}} \equiv \frac{\theta_0 u_*^2}{\kappa g \theta_*} \tag{5.18}$$

In Eq. (5.16), $\phi_m(\zeta), \phi_h(\zeta)$ are non-dimensional "universal" functions to be determined with experimental data. Moreover, we can write the expression for the Richardson number — which compares buoyancy and shear contributions — in terms of these universal functions,

$$\text{Ri} \equiv \frac{g\partial \ell n\bar{\theta}/\partial z}{(\partial \bar{u}/\partial z)^2} = \zeta\frac{\phi_h}{\phi_m^2} \tag{5.19}$$

Integration of Eq. (5.16) provides the mean velocity and temperature ($\bar{u}, \bar{\theta}$, respectively) in the surface layer where the vertical fluxes are constant. To perform these integrations, we must set boundary conditions. These are taken at $\zeta_0 = z_0/L$, where z_0 is referred to as the roughness length that depends on the physical properties of the surface. The boundary conditions are $\bar{u}_0 = \bar{u}(z_0) = 0$, $\bar{\theta}_0 = \bar{\theta}(z_0)$. Thus, integration of the expression in Eq. (5.16) from z_0 up to a height z within the surface layer gives

$$\bar{u} = \frac{u_*}{\kappa}\Phi_m(\zeta) \tag{5.20}$$

$$\bar{\theta} - \bar{\theta}_0 = \frac{\theta_*}{\kappa}\Phi_h(\zeta) \tag{5.21}$$

where

$$\Phi_m = \int_{\zeta_0}^{\zeta}\frac{1}{\zeta}\phi_m(\zeta)d\zeta \quad \Phi_h = \int_{\zeta_0}^{\zeta}\frac{1}{\zeta}\phi_h(\zeta)d\zeta \tag{5.22}$$

where $\Phi_m(\zeta_0) = \Phi_h(\zeta_0) = 0$ by definition. Note that $\bar{\theta}_0$ in Eq. (5.21) and θ_0 in Eq. (5.18) are not identical; the former is the mean of θ at $z = z_0$ while the latter comes from the anelastic approximation. Equations (5.16)–(5.22) allow for the calculation of $u_*^2 \equiv -\overline{w'u'}$ and of $u_*\theta_* \equiv -\overline{w'\theta'}$, which are the fluxes of momentum and heat.

In the following we replace $\bar{u}$ by the horizontal velocity vector $\bar{\mathbf{v}}$ in order to provide a more general formulation of the surface eddy diffusion and transfer coefficients. Using this notation, the fluxes at the surface are given by

$$\overline{w'\mathbf{v}'} = -C_D|\bar{\mathbf{v}}|\bar{\mathbf{v}} \tag{5.23}$$

$$\overline{w'\theta'} = -C_H|\bar{\mathbf{v}}|(\bar{\theta} - \theta_s) \tag{5.24}$$

where the subscript "s" indicates surface conditions and,

$$C_D = \left(\frac{\kappa}{\Phi_m}\right)^2 \quad C_H = \frac{\kappa^2}{\Phi_m\Phi_H} \tag{5.25}$$

The coefficients C_D, C_H can be obtained from curves that are parametric in z/z_0 and function of the bulk Richardson number (Louis, 1979; Louis et al., 1982), which is defined as

$$\mathrm{Ri}_B \equiv g\frac{\bar{\theta} - \theta_s}{\theta_0}\frac{z}{\mathbf{v}^2} = \zeta\frac{\Phi_h}{\Phi_m^2} \tag{5.26}$$

These expressions are "local" in the sense that the vertical profile of turbulent diffusivity depends on the local vertical shear and static stability.

For the vertical flux of virtual potential temperature and moisture, the following expressions are usually assumed,

$$\overline{w'\theta'_v} = -C_H|\bar{\mathbf{v}}|(\bar{\theta}_v - \theta_{vs}) \tag{5.27}$$

$$\overline{w'q'} = -C_E|\bar{\mathbf{v}}|(\bar{q} - q_S^*) \tag{5.28}$$

where C_E is the transfer coefficient for evaporation. In some applications two or more of the transfer coefficients are given the same value.

The arguments in this subsection have provided a rationale for use of the bulk aerodynamic formula in AGCM on the basis of the equations for the turbulence flow above a surface. The way in which these formula are used in AGCMs will be addressed in the following sections.

5.5.2   The turbulent mixing coefficients as a function of Ri

For the turbulent flux of a flow variable $A(P,t)$ we write

$$\overline{w'A'} \equiv -K\frac{\partial\bar{A}}{\partial z} \tag{5.29}$$

where K is the eddy mixing or transfer coefficient. Equation (5.29) is a closure assumption that gives the turbulent flux as a function of the mean flow. Using the two expressions in Eq. (5.16), we can write for the turbulent vertical flux of momentum,

$$\overline{w'u'} \equiv -u_*^2 = -\left(\frac{\kappa z}{\phi_m}\right)^2 \left|\frac{\partial\bar{u}}{\partial z}\right|\frac{\partial\bar{u}}{\partial z} = -K_m\frac{\partial\bar{u}}{\partial z} \tag{5.30}$$

Similarly, for the turbulent vertical flux of heat,

$$\overline{w'\theta'} = -u_*\theta_* = -\frac{\kappa z}{\phi_m}\left|\frac{\partial\bar{u}}{\partial z}\right|\frac{\kappa z}{\phi_h}\frac{\partial\bar{\theta}}{\partial z} = -\frac{(\kappa z)^2}{\phi_m\phi_h}\left|\frac{\partial\bar{u}}{\partial z}\right|\frac{\partial\bar{\theta}}{\partial z} = -K_h\frac{\partial\bar{\theta}}{\partial z} \tag{5.31}$$

Therefore, the mixing coefficients can be written in the following way,

$$K_m = \ell_0^2 f_m(\text{Ri})\left|\frac{\partial\bar{u}}{\partial z}\right| \qquad K_h = \ell_0^2 f_h(\text{Ri})\left|\frac{\partial\bar{u}}{\partial z}\right| \tag{5.32}$$

where

$$\ell_0 = \kappa z, \quad f_m(\text{Ri}) = \frac{1}{\phi_m^2}, \quad f_h(\text{Ri}) = \frac{1}{\phi_m\phi_h} \tag{5.33}$$

and ℓ_0 is referred to as Prandtl's mixing length (see the Appendix for a justification of this notation).

The procedure to obtain the fluxes of momentum and heat at height z for a given mean flow $\bar{u}(z), \bar{\theta}(z)$, consists of the following steps: (1) Compute the value of the Richardson number using Eq. (5.19) and use graphical representations of Ri(ζ) [e.g., Businger et al, 1971] to obtain the value of ζ, (2) Solve for ϕ_h, ϕ_m using the distributions of $\phi_h(\zeta), \phi_m(\zeta)$ given by experimental data, (3) Find u_*, θ_* from Eq. (5.16), and obtain the fluxes from Eq. (5.17).

For a moist atmosphere, θ can be formally replaced by the virtual potential temperature θ_v, which is defined by

$$\theta_v = (1 + 0.6q)\theta \tag{5.34}$$

where $R_v/R - 1 = 0.6$. In this case, the expression for the vertical flux of virtual potential temperature becomes

$$\overline{w'\theta_v'} = -K_h\frac{\partial\bar{\theta}}{\partial z} \tag{5.35}$$

5.6   Planetary boundary layer structure and height

We examined in Sec. 5.5 the structure of the surface layer, which depends fundamentally on the physical processes that are primarily responsible for generating the turbulence. Turbulent eddies near the Earth's surface have many time scales. Some eddies associated with convective cells can eventually extend beyond the surface layer until they lose buoyancy and even continue upwards by inertia and coexist with others at smaller scales. Conceptually, the small, quasi-isotropic eddies are locally diffusive while the larger, non-isotropic and penetrative have non-local effects. Above the surface layer is the outer boundary layer, which also depends fundamentally on the turbulent processes. The ensemble of surface and outer boundary layer form the planetary boundary layer (PBL). Further above, and through a complex interface representing the PBL top, is the free atmosphere.

A schematic of the PBL structure at different stages in the diurnal cycle is shown in Fig. 5.4. The surface layer is present at all times with variable thickness and showing the emergence of convective eddies in the early afternoon. The outer PBL becomes thicker during the day and collapses at night. In locations around the world, deep convection can develop and the associated mass flux (cumulus mass flux) can detrain air from the PBL to the free atmosphere. The evolution portrayed in both panels of Fig. 5.4 is representative of the situation over tropical lands. Over oceans with cold sea surface temperature (SST) such as in the eastern parts of the tropical Pacific and Atlantic (see Ch. 1) the clouds at the PBL top persist during the diurnal cycle and there is no penetrative convection.

Analyses of vertical soundings from observational campaigns have shown the existence of sharp transitions associated with the PBL in many parts of the world with climate significance [Seidel et al., 2012]. A review of methods to

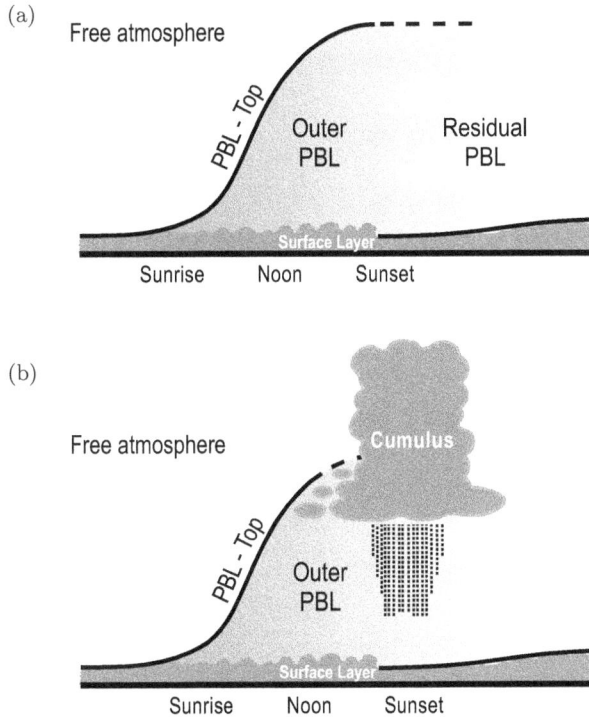

Figure 5.4.   Schematic of the PBL structure at different stages in the diurnal cycle. (Courtesy of C. S. Konor.)

estimate the PBL height is in Seibert et al. [2000]. In the following we present a few examples of such methods.

Some methods identify the PBL top with locations in which the decrease with height of air temperature is interrupted by a shallow layer within which the temperature increases (inversion layer) and above which the atmosphere is stably stratified. In Heffter [1980] the temperature inversion is identified as the base in the range of heights at which the potential temperature difference between the base and top satisfies

$$\Delta\theta/\Delta z \geq 0.5\,\mathrm{K}/100\,\mathrm{m} \quad \text{for} \quad \Delta\theta > 2\,\mathrm{K} \tag{5.36}$$

Other methods are based on surface fluxes of momentum and heat together with scaling parameters, such as the Monin–Obukhov length L (Eq. (5.18)). For near neutral conditions, the inversion height can be obtained from the following relationship [Blackadar and Tennekes, 1968],

$$h = \frac{c_n u_*}{|f|} \quad \text{when} \quad \left|\frac{u_*}{fL}\right| < 4 \tag{5.37}$$

where the constant c_n was set to 0.2, and a minimum value of f is assumed to correspond to $20°$ latitude. For stable conditions, the expressions are [Zilitinkevich, 1972; Zilitinkevich et al., 2002],

$$h = c_s \left(\frac{u_* L}{|f|}\right)^{1/2} \quad \text{when} \quad \left|\frac{u_*}{fL}\right| > 4 \tag{5.38}$$

where c_s was set to 0.4. Alternatively, Nieuwstadt (1981) proposed

$$h = \frac{c_1 u_*/|f|}{1 + c_2 h/L} \tag{5.39}$$

which reduces to Eq. (5.38) for the neutral and stable cases when $L \to \infty$ and $L \to 0$, respectively.

Estimates of the PBL height using the Richardson number have also been proposed. The procedure is based on searching the vertical sounding for the location at which the Richardson number in Eq. (5.19) drops below a critical value:

$$\mathrm{Ri} < \mathrm{Ri}_c \tag{5.40}$$

In theoretical or laboratory work, Ri_c is around 0.25. In practice and according to the data available, Ri_c can be values up to 0.55 [Straume et al., 1998].

For stable conditions, a definition based on the bulk Richardson number defined in Eq. (5.26) is

$$h = \text{Ri}_{Bc} \frac{\mathbf{v}^2}{(g/\theta_{v0})(\bar{\theta}_{vh} - \bar{\theta}_{vS})} \tag{5.41}$$

where Ri_{Bc} can be 0.51 [Mahrt, 1981] or 0.33 [Wetzel, 1982].

We emphasize that the PBL depth does not only depend on factors below the PBL top. Some of these factors are stratification of the free atmosphere, large-scale subsidence, whether or not the PBL is cloud topped, and the past history of the PBL as in diurnal cycles. It is important to recognize that the PBL is not independent of the atmosphere above. Modeling of the PBL, therefore, is both challenging and is crucial for climate simulations with models of the atmosphere.

5.7 Turbulence above the surface layer

In the PBL above the surface layer (see Fig. 5.4), the treatment of turbulent processes recognizes two conceptual approaches according to the importance of the role assigned to turbulent eddies. One approach emphasizes the small eddies, which are assumed to be quasi-isotropic and diffusive. One methodology within this approach is based on Richardson number-dependent, K-closure models (sometimes called first-order closure models). Similar to Eq. (5.27) for the surface layer we write

$$K_m = \ell^2 f_m(\text{Ri}) \left| \frac{\partial \bar{\mathbf{v}}}{\partial z} \right|, \quad K_h = \ell^2 f_h(\text{Ri}) \left| \frac{\partial \bar{\mathbf{v}}}{\partial z} \right| \tag{5.42}$$

This approach is further discussed in Sec. 5.5.2.

The other approach to the treatment of turbulent processes in the outer boundary layer emphasizes the role of large eddies, which are non-isotropic and non-local. One methodology that follows this approach is based on considering the outer PBL (see Fig. 5.4) as a mixed layer, in which conservative variables are assumed to be homogenized by turbulence processes. Schematics of the vertical distribution of a conservative variable (referred to as θ) are given in Fig. 5.5 for cases in which either no clouds are present or clouds are present in the PBL. The latter situation occurs over the eastern part of the southern subtropical oceans. In either case, turbulence contributes to mass entrainment from the free atmosphere. When clouds are present, radiative cooling at the top also contributes to the mass entrainment. This approach is further discussed in Sec. 5.5.3.

Atmospheric turbulence is a very active area of research, and other approaches have been advanced. In the Richardson number-dependent, K-closure models (Eq. 5.37), the second order moments $\overline{w'u'}, \overline{w'\theta'}$ are written in terms of mean flow properties. However, the equations governing these second moments involve triple correlations of the type $\overline{w'u'u'}, \overline{u'w'\theta'}$, etc.

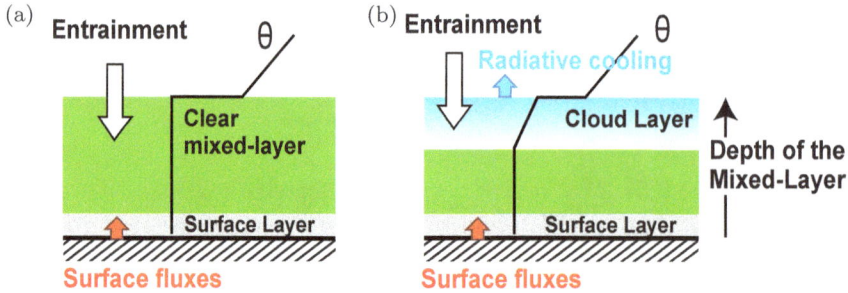

Figure 5.5.   Schematics of the vertical distribution of a conservative variable in the mixed-layer approach to treat turbulent processes in the outer PBL. (Courtesy of C. S. Konor.)

Mellor and Yamada (1974, 1982) proposed several ways to add closure assumptions for the third-order moments in terms of second-order moments such that the prognostic equations for the latter would not have higher order terms. These different ways are referred to as a "hierarchy". A particularly popular member of the hierarchy is the 2.5 level, in which there is only one prognostic variable, which is the turbulent kinetic energy (TKE, also referred to as e, in the literature). That is, the diffusion coefficients are written in terms of e. The reader is referred to the papers mentioned above in this paragraph for further information.

More recently, Bretherton and Park [2009] presented a parameterization of turbulence that emphasizes moist processes and is based on down-gradient diffusion of moist conserved variables. One motivation for this development was to improve the simulation of PBL clouds (see Figs. 5.4 and 5.5) which is challenging for AGCMs (see Ch. 7). Another relevant development is the Cloud Layers Unified by Binormals (CLUBB), which unifies the PBL and shallow convection parameterizations with a higher-order turbulence closure [Bogenschutz et al., 2012, 2013]. In a version of CLUBB [Bogenschutz et al., 2012] the following correlations are predicted $\overline{u'^2}, \overline{v'^2}, \overline{\theta_l'^2}, \overline{q_t'^2}, \overline{\theta_l'q_t'}, \overline{w'q_t'}, \overline{w'^2}, \overline{w'^3}$, where the subscripts "$l$" and "$t$" denote liquid and total, respectively.

5.8   Examples of PBL parameterizations

PBL parameterizations are expected to provide information on surface and vertical fluxes of momentum (stress), heat and moisture, PBL depth, as well as cloud cover. The analysis of PBL turbulence in the previous sections has provided a rationale for use of the bulk aerodynamic formula to estimate the surface fluxes of momentum, heat and water from the surface into the atmosphere. However, a direct transplant of the analysis results in AGCMs poses conceptual difficulties. One of these difficulties is how to estimate in AGCMs the roughness length z_0 (see Sec. 5.5.1). In the early stages AGCM development, z_0 was assigned different values over sea, land and sea-ice. Contemporary AGCMs may prescribe geographically and seasonally varying distribution of z_0 or even predict them using algorithms that take into account the state of the sea surface (e.g., the surface wave field) and the vegetation type (e.g., forest or cultivate land). Other difficulties result from the decision on how to select the values of velocity, temperature and humidity in Eqs. (5.23), (5.24), (5.27) and (5.28). These expressions apply to the surface layer, which is about 100 m deep. However, the vertical resolution of the models is not enough to resolve the velocity profiles at such level of detail. Different ways to address this issue have been selected by different applications.

In the Zebiak and Cane [1987; ZC87] model, for example, the aerodynamic formula is used to compute the wind stress at the ocean surface (Eq. (4.58) in Ch. 4). ZC87 is a seasonal anomaly model and the surface velocity $\mathbf{v}$ is assumed to be the climatological seasonal surface wind vector as obtained from some observationally-based dataset plus the perturbation velocity $\mathbf{v}'$ corresponding to the model's lowest level. Furthermore, C_D is assumed to be constant. In the Mintz–Arakawa model mentioned in Sec. 5.2, the surface velocity is obtained by extrapolation using the velocities available at the two model levels, the lowest being at about 2.5 km above the surface, multiplied by a "tuning" constant. For more recent AGCMs with multiple layers the extrapolation to the surface seems more justifiable, especially if the layers are within the PBL. This may not be possible at all times because the PBL may collapse at night and the winds in the PBL itself may have a fine structure over complex terrain (see Fig. 5.4). Therefore, we will consider that the choice of L belongs to the PBL parameterization itself.

Therefore, it is more appropriate to view the aerodynamic formula in climate models as diagnostic relationships for the surface layer values of the mean field from the fluxes determined by the processes in the entire atmospheric column. The applicability of this concept in the PBL parameterization context is discussed in two examples corresponding to well-known AGCMs in the following sections.

5.8.1 *The PBL parameterization in CAM4*

The Community Atmospheric Model version 4 (CAM4) has been used by research groups around the world. The PBL parameterization in CAM4 is the same as the one in the previous version CAM3. A scientific description of CAM4 is in NCAR Technical Note: http://www.cesm.ucar.edu/models/ccsm4.0/cam/ docs/description/cam4_desc.pdf. Although this is not the most recent version of CAM, it offers a simple but yet complex framework to discuss outstanding issues in implementing the evaluation of surface fluxes.

In view of our interest on atmosphere-ocean interactions, we concentrate on the atmosphere over the oceans. In CAM3/4/5 the transfer coefficients

C_D, C_H, C_E at a height z_A over the ocean are given by

$$C_{(D,H,E)} = \kappa^2 \left[\ell n \left(\frac{z_A}{z_{0m}} \right) - \psi_m \right]^{-1} \left[\ell n \left(\frac{z_A}{z_{0(m,e,h)}} \right) - \psi_{(m,s,s)} \right]^{-1} \quad (5.43)$$

where $z_{0(m,e,h)}$ are the roughness length of momentum, heat and evaporation, respectively. The profiles of ψ are given as function of a stability parameter that depends on the von Karman constant κ, z_A, the difference between the velocities, temperature, and water vapor mixing ratio at level z_A and the surface (where the velocity is zero), and of the coefficients C_D, C_H, C_E themselves. The profiles are different for stable and unstable conditions. The corresponding expressions are, therefore, implicit in the coefficients and must be solved by iteration.

In the free atmosphere, no distinction is made between turbulent diffusion coefficients for momentum, heat, or water vapor. These are given by

$$K_C = \ell_C^2 f_C(\text{Ri}) \left| \frac{\partial \bar{\mathbf{v}}}{\partial z} \right| \quad (5.44)$$

in which, following Blackadar [1962], ℓ_C is given by

$$\frac{1}{\ell_C} = \frac{1}{kz} + \frac{1}{\lambda_\infty} \quad (5.45)$$

where λ_∞ is a prescribed value (30–100 m), and

$$f_C(\text{Ri}) = \begin{cases} \dfrac{1}{1 + b\,\text{Ri}(1 + b'\text{Ri})} & \text{Ri} > 0 \\ (1 - c\,\text{Ri})^{1/2} & \text{Ri} < 0 \end{cases} \quad (5.46)$$

where $b = 10$, $b' = 8$, $c = 18$. In this formulation, turbulent fluxes are just locally related to the mean gradient. Redistribution of turbulence due to transport or diffusion are not considered. These assumptions are reasonable when the length scale of the largest turbulent eddies is smaller than the size of the domain over which the turbulence extends.

In running the model, the diffusivities in the free atmosphere are computed first at all levels. However, the assumption of locality does not hold for a PBL in unstable and convective conditions. In such cases, turbulent eddies may have a similar size to the depth of the boundary layer and the flux can be opposed to the local gradient. To account for this, CAM4 estimates a height h for the PBL. The relationships used to estimate h depend on the height of the underlying surface, the horizontal velocities and temperature at the lowest model level, a critical Richardson number, and other parameters. At locations within the PBL, diffusivities are replaced by the values computed in the following way [Troen and Mahrt, 1986; Holtslag and Boville, 1993],

$$\overline{w'A'} = -K_A \left(\frac{\partial \bar{A}}{\partial z} - \gamma_A \right) \quad (5.47)$$

where the non-local diffusion coefficient is given by

$$K_A = k w_t z \left(1 - \frac{z}{h}\right)^2 \tag{5.48}$$

in which w_t is a turbulent velocity scale and h is the boundary layer height. In Eq. (5.47), γ_A represents non-local influences on the mixing by turbulence [Deardorff, 1972], which is neglected under stable conditions. This scheme is optimized for simulation of dry convective and nocturnal boundary layers over land. Eddy diffusivity can also have a prescribed vertical profile, which is controlled by the surface heat flux [e.g., Troen and Mahrt, 1986; Holtslag et al., 1990; Holtslag and Boville, 1993]. In these schemes, the height of the PBL top is diagnostically determined without explicitly considering the mass entrainment through the PBL top. Bretherton and Park (2009) have revised the Holton and Boville [1993] scheme in order to provide a more realistic treatment of stratocumulus-topped PBLs.

5.8.2 *The PBL parameterization in the UCLA AGCM*

This is an example of PBL parameterization following a bulk approach as in Deardorff [1972]. If a flow variable is well-mixed (in the vertical), then the left-hand side in Eq. (5.13) is independent of height and the turbulent fluxes have to vary linearly with height. A schematic of the profiles of different variables in the PBL is given in Fig. 5.6.

In the UCLA AGCM framework, the PBL has a variable depth and is assigned to the lowest layer of the model [Suarez et al., 1983]. This is implemented by a generalized vertical coordinate, in which the PBL top is the first coordinate surface from the ground up [Suarez et al., 1983]. The major advantage of using such a coordinate is that PBL properties are expected

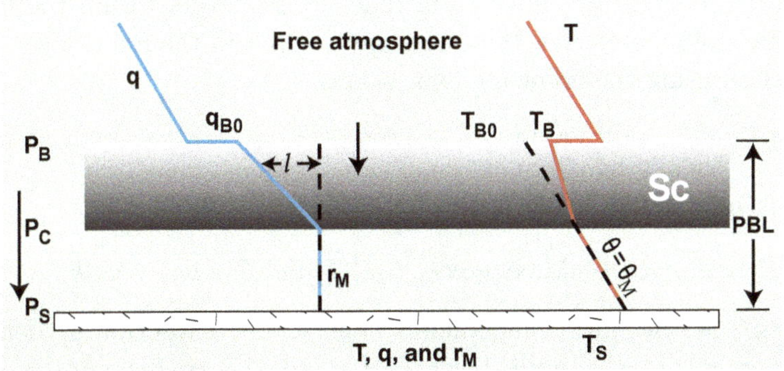

Figure 5.6.   Schematic of profiles of prognostic variables in a PBL parameterization used in the UCLA AGCM. (Courtesy of C. S. Konor.)

to be "similar" along a coordinate surface. Clouds form within the PBL at heights above the lifting condensation level and the PBL top. This configuration facilitates the formulation of processes concentrated near the PBL top. The depth h of this PBL/AGCM's lowest layer is predicted through the mass budget equation including the PBL-top entrainment that depends on whether the PBL is cloud-free or cloud-topped, cumulus mass flux (given by the convection parameterization), and horizontal mass convergence within the PBL (see Fig. 5.5). Stratification at cloud top affects the buoyancy flux and hence the entrainment rate [Stevens, 2002]. When the entrainment and cumulus mass flux become zero, the PBL-top becomes a material surface, keeping the PBL air separated from the free atmosphere air.

The expressions that determine the surface fluxes are similar to the aerodynamic formula but the mean values over the entire PBL are used instead of the values at a level L above the surface as in Eq. (5.1)

$$
\begin{aligned}
(\mathbf{F_v})_S &\equiv -\rho_S C_M |\mathbf{v}_M| \mathbf{v}_M \\
(F_\theta)_S &\equiv \rho_S u_* C_\theta (\theta_S - \theta_M) \\
(F_q)_S &\equiv \rho_S u_* C_\theta \beta [q^*(T_S, p_S) - r_M] \\
u_*{}^2 &= |\mathbf{F_v}|/\rho_S
\end{aligned}
\qquad (5.49)
$$

where β is the coefficient of potential evaporation (see Sec. 5.10). The coefficients C_M, C_θ are given by analytic expressions that are parametric in z_0 and function of the bulk Richardson number defined by

$$
\text{Ri} \equiv g \frac{\bar{\theta}_m - \theta_S}{\theta_{0m}} \frac{h}{|\bar{\mathbf{v}}|_m^2}
\qquad (5.50)
$$

A true mixed-layer approach, however, excludes the vertical shear and associated generation of turbulent kinetic energy within the PBL. It also sets the bottom level of cumulus clouds and the top level of stratocumulus clouds (Sc). Konor and Arakawa [2005, 2009] suggested that the traditional mixed layer methodology could be extended to include several layers within the PBL. The surface fluxes of momentum, heat, and water vapor in the multi-layer approach are computed using the following expressions,

$$
\begin{aligned}
(\mathbf{F_v})_S &\equiv -\rho_S C_U C_U \text{Max}(\alpha_1 |\mathbf{v}_L|, \beta_1 e_{PBL}^{1/2}) \mathbf{v}_L \\
(F_\theta)_S &\equiv \rho_S C_U C_T \text{Max}(\alpha_2 |\mathbf{v}_L|, \beta_2 e_{PBL}^{1/2})(\theta_S - \theta_L) \\
(F_q)_S &\equiv \rho_S C_U C_T \text{Max}(\alpha_2 |\mathbf{v}_L|, \beta_2 e_{PBL}^{1/2})[q^*(T_S, p_S) - q_L] k
\end{aligned}
\qquad (5.51)
$$

where e_{PBL} is the turbulent kinetic energy, the subscript L denotes the lowest layer within the multi-layer PBL, and the coefficients $\alpha_1, \alpha_2, \beta_1, \beta_2$ are empirically determined constants. The new formulation has produced encouraging results but has not been fully implemented in an operational model.

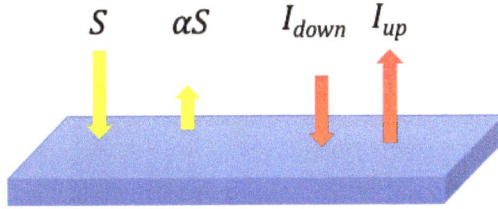

Figure 5.7.   Radiative fluxes at the Earth's surface. Yellow arrows indicate solar radiation and red arrows are for terrestrial radiation.

5.9   Radiation fluxes at the atmosphere lower boundary

In AGCMs, the complex calculation of radiative fluxes is done in a highly specialized module called the radiation parameterization. The input to the radiation scheme consists of the surface properties and profiles of atmospheric properties, and the output consists of profiles of broadband longwave and shortwave fluxes. Among the atmospheric properties are the optical properties of gases, clouds and aerosols [Hogan and Bozzo, 2016]. Among the surface properties a very important parameter is the albedo (α), which indicates the fraction of solar radiation that is reflected back from the surface (Fig. 5.7).

The value of α can have strong spatial and temporal variations. The albedo of the ocean surface is about 0.06 (which may depend on the state of the sea) while it can be about one order of magnitude larger over bare sea-ice. The calculation of radiative fluxes at the sea-ice surface, therefore, must take into account that a large fraction of incident solar radiation may be reflected back to space depending on the relative surface distributions of water and ice, and on whether the ocean surface temperature is at its freezing level, which is also a function of the water composition, or whether the ice is covered by snow and this is fresh or it has aged. We have discussed in Ch.1 an important feedback process that involves the albedo. On land, the reflectivity of sunlight is highly dependent on the cover type.

5.10   Surface fluxes over sea-ice and land surfaces

The importance of recognizing the difference in surface conditions — particularly sea-ice and land — has been recognized since the earliest AGCMs were developed. However, the importance of this recognition goes well beyond choosing different values of z_0 in the calculation of the transfer coefficients for the turbulent fluxes. One additional aspect is to take into account the different ways in which solar radiation is reflected; as addressed in Sec. 5.9. In this subsection we look at the turbulent fluxes above sea-ice and land surface.

In a broad sense, both sea ice and land are treated as slabs located in between the atmosphere and either the sea-ice bottom in contact with the ocean or the depth beyond which no moisture is expected to make its way up to the land surface. The slabs can be single or multilayer, according to the algorithm used. The sea-ice model is generally incorporated in the OGCM; some more details are given in Sec. 6.2.3. At this point we just mention the calculation of surface temperature for the sea-ice. The value of this temperature is obtained by solving jointly the equations for the surface energy budget and the vertical diffusion of heat within the slab from freezing temperatures at its lower boundary. In some implementations of the coupling between atmosphere and sea-ice (or land), the vertical diffusion of heat is made from the top of the atmosphere to the bottom of the ice or land model (see Sec. 8.2.4).

The slab representing land had prescribed moisture conditions in early AGCMs. In this configuration, dry or wet conditions are artificially maintained whatever the atmospheric circulation is. Such a technique may produce reasonable mean evaporation fields, but it necessarily precludes land-atmosphere feedback processes, some of which lie at the heart of continental climate sensitivity. For this reason, interactive land surface models (LSMs) were developed. The simplest is the so-called "bucket" model of Manabe [1969], which allows the water level in a soil moisture reservoir to increase during precipitation events and to decrease as the water evaporates. Because the evaporation efficiency varies with the water level in the reservoir, rainy periods lead to high evaporation rates, and droughts lead to low rates.

Evaporation from the land surface is often assumed as

$$E = \beta E_p \qquad (5.52)$$

where E_p is the rate of potential evaporation (as expected when the ground is wet) and the coefficient β is an aridity index, which depends on soil conditions, such as type, vegetation cover, etc. Plants transpire moisture, and the surface moisture flux over vegetated land is due to evapotranspiration. Prediction of β requires specialized algorithms that are part of LSMs.

A more complex representation of climate-vegetation biophysical processes is provided by the soil-vegetation-atmosphere-transfer (SVAT) model [Sellers et al., 1996; Dickinson et al., 1998]. This includes photosynthesis-transpiration physics and considers the spatial heterogeneity in subgrid features in surface characteristics [e.g., Entekhabi and Eagleson, 1989; Famiglietti and Wood, 1991; Avissar, 1992; Koster and Suarez, 1992; Ducoudre et al., 1993; Seth et al., 1994; Quinn et al., 1995; Stieglitz et al., 1997].

5.11   Examples of AGCMs coupled to simple ocean models

As stated in Ch. 2, we refer to an AGCM coupled to a simple ocean model as a hybrid coupled model of Type 1 (HCM1).

5.11.1 *AGCMs coupled to swamp ocean models*

A swamp ocean (SWO) is a body of water that is motionless and cannot store heat (i.e., has zero heat capacity). SSTs are either prescribed or computed from surface energy balance only. The ocean surface provides an infinite source of water (latent heat) for the atmosphere above. Such a simpler type coupled model is useful for studies on climate feedbacks, especially those that involve fast adjustment processes.

The first attempt to use an HCM1 configured as an AGCM coupled to a SWO was carried out by Manabe et al. [1975]. The goal was to achieve a better understanding of the role of ocean currents on climate. These authors compared the oceanic heat transport computed by a coupled ocean-atmosphere model consisting of the same AGCM coupled to either an oceanic general circulation model (OGCM, see Ch. 6) or to a SWO. They concluded that the total poleward energy transport is affected little by the energy transport by ocean currents. This is because the atmospheric energy transport was decreased by approximately the same magnitude as the oceanic energy transport in the presence of ocean dynamics.

5.11.2 *AGCMs coupled to slab ocean models*

An SOM (see Ch. 2) is a motionless body of water with constant depth (i.e., has non-zero heat capacity). The slab can be interpreted as an approximation of a well-mixed ocean mixed layer. The lack of dynamical effects means the absence of heat transport by the ocean, and hence AGCMs coupled to SOMs have major difficulties in simulating the mean and seasonally varying distributions of SST and sea ice. Therefore, corrections must be artificially incorporated in such HCM1s [e.g., Bitz et al., 2012].

The SST equation in an SOM is written as

$$\rho_0 c_p h \frac{\partial SST}{\partial t} = F_{net} + Q_{flux} \qquad (5.53)$$

where ρ_0 is the density of seawater used, c_p is the specific heat of seawater at constant pressure, h is slab (mixed later) depth, and F_{net} is net heat flux at the ocean interface with the atmosphere (provided by the AGCM) and other systems such as sea-ice. Q_{flux} is a flux correction term that is obtained by applying the same equation except that the other two terms are obtained from a long-term climatology by the same AGCM to be coupled. Use of such correction results in a reasonable seasonal distribution of SSTs and sea-ice cover. An additional relaxation term to climatology may be also necessary

Washington and Meehl [1984] used an HCM1 configured as an AGCM coupled to an SOM in climate sensitivity experiments to doubling and quadrupling the present amount of atmospheric carbon dioxide (CO_2), with both fixed and computed clouds. They found that the global-mean surface air temperature in the doubled (quadrupled) CO_2 experiment increased by 1.3°C (2.7°C) for the fixed clouds and 1.3°C (3.4°C) for the computed clouds, and the stratosphere cooled by 6°C (11°C).

An AGCM/SOM configuration was applied to assess the equilibrium climate sensitivity to CO_2 doubling. In Danabasoglu and Gent [2009], the AGCM is the low-resolution version of the Community Climate System Model, version 3 (CSM3). The globally averaged equilibrium surface temperature response to $2 \times CO_2$ produced by the AGCM/SOM is about 0.14°C lower than that obtained by the AGCM coupled to a full ocean general circulation model (OGCM, see Ch. 6). The conclusion was that the simpler system could give a good estimate of the equilibrium climate sensitivity as compared to the more complex system. Furthermore, Gregory et al. [2004] showed that the climate feedback parameters for evaluating the radiative forcing associated with increasing CO_2 estimated from an AGCM/SOM system were consistent with those obtained in an AGCM/OGCM system.

Using an AGCM coupled to an SOM in which the domain is set as the Atlantic basin, Trzaska et al. (2007) found that the leading simulated empirical coupled mode in the tropics resembles the zonal mode, despite the lack of ocean dynamics. This mode is associated with baroclinic atmospheric anomalies in the tropics and a Rossby wave train extending to the extratropics, suggesting an atmospheric response to tropical SST forcing.

In hybrid coupled models, SOMs have also been used in Aquaplanet configurations (see Ch. 2 for a definition). For example, Seo et al. [2014] employed an AGCM/Aquaplanet SOM to examine the efficiency of surface thermal forcing on shifting the meridional position of the simulated Intertropical Convergence Zone (ITCZ). They found that climate variations in the extratropics can influence important features of the tropical circulation. In particular, extratropical thermal forcing can shift the ITCZ even more than tropical thermal forcing.

The extent to which the biases in ocean mixed layer depth affect those in SST was tested by using an AGCM coupled to an SOM. Z. Wang et al. [2019] reported that if the fixed mixed layer depth in this model is reduced by a factor of two, then the annual mean SSTs experience a strong cooling (more than 4 K) over the tropics and a warming over the high latitudes. Interestingly, the annual-mean net surface heat flux does not change because of the equilibrium energy budget constraint of the SOM, while the annual cycle does change. Thus, the rectification of asymmetrical changes in annual cycle of SST driven by change in the fixed mixed layer depth of the SOM led to changes in the annual-mean SSTs.

The Madden–Julian oscillation (MJO) is a well-known feature of the intraseasonal variability of the atmosphere occurring mainly over the tropics.

Although the main processes at work in the MJO can be addressed without coupling to the ocean, the importance of oceanic feedback in MJO theory has been receiving increased interest [DeMott et al., 2015]. Maloney and Sobel [2004] carried out a sensitivity experiment of MJO to the mixed layer depth using an AGCM coupled to a SOM. They found that the MJO variability in precipitation obtained with a fixed mixed layer depth was enhanced relative to a fixed-SST simulation, mainly via a wind-induced surface heat exchange feedback.

ENSO is known to be one of the strongest air-sea coupled phenomenon. However, Dommenget [2010] showed that the atmospheric model simulations coupled to a slab ocean model could produce El Niño type of SST variability. The evolution of SST resembles an SST-mode of El Niño [Neelin, 1991], which is induced by interactions between atmosphere and ocean dynamics at the surface layer, but the physical processes at work are not the same because SOM has no ocean dynamics. The evolution of SST was attributed to thermo-dynamical feedback between atmosphere and ocean, especially through cloud-shortwave-SST feedback, with the heat capacity of the upper ocean playing a role in the damping process.

5.12 Appendix A: Classical mixing length theory

The classical mixing length theory applies to turbulent flows in the neutral case with no stratification of θ. In this case we can define a velocity scale that is given by the frictional velocity due to the approximate constancy of fluxes in the layer. In these cases, it is reasonable to assume that u_* defined in Eq. (5.16) is the only parameter that characterizes the vertical shear of the mean velocity, that is,

$$\frac{\kappa z}{u_*}\frac{\partial \bar{u}}{\partial z} = 1 \tag{5A.1}$$

where κ is the von Karman constant. Integration of Eq. (5A.1) with respect to z gives

$$\bar{u} = \frac{u_*}{\kappa} \ell n \left(\frac{z}{z_0}\right) \tag{5A.2}$$

Note that $u(z_0) = 0$.

To obtain "closure" in turbulence equations we must add relationships between mean and eddy components. A simple relationship can be written by assuming that for an arbitrary variable $A(z, t)$ the following relationship holds,

$$\overline{w'A'} = -K\frac{\partial \bar{A}}{\partial z} \tag{5A.3}$$

where $K(P, t)$ is an eddy diffusion coefficient, which becomes a property of the flow itself. The classical mixing length theory [Prandtl, 1925] assumes

$$K = \ell^2 \left| \frac{\partial \bar{\mathbf{v}}}{\partial z} \right| \tag{5A.4}$$

This is consistent with assuming that if a parcel is displaced vertically a distance proportional to $\ell(P, t)$, then the eddy vertical velocity varies linearly with the modulus of the horizontal eddy velocity and the other eddy variables linearly in z.

The finding and estimation of $\ell(z)$ have motivated extensive experimental work. Let us consider a location close to the surface. Here it is reasonable to assume that $\ell \cong \alpha z$ and $|\partial \bar{\mathbf{v}}/\partial z| \cong \bar{\mathbf{v}}/z$. Moreover, taking $A = \bar{\mathbf{v}}$, we can calculate at the surface denoted by the subscript "S",

$$(\overline{\rho_0 w' \mathbf{v}'})_S = -\rho_0 C_D |\bar{\mathbf{v}}| \bar{\mathbf{v}} \tag{5A.5}$$

This expression $(C_D = \alpha^2)$ resembles that provided by the bulk aerodynamic method to obtain the surface stress. C_D can be identified as the drag coefficient.

Chapter 6

OGCMs Coupled to Simpler
Atmospheric Models

6.1 Introduction

This chapter is dedicated to an important family of coupled atmosphere-ocean systems in which the oceanic component is a general circulation model (OGCM), while the atmospheric component is a simpler model. OGCMs synthesize a number of mutually interacting processes in the oceans that are deemed as essential for climate. Oceans store and transport enormous amounts of heat. Hence, OGCMs are essential tools for studies on climate variability and change. A brief description of OGCM fundamentals is presented; the simpler models of the atmosphere have already been mentioned in previous chapters. As stated in Ch. 2, we refer to an OGCM coupled to a simple atmospheric model as a hybrid coupled model of Type 2 (HCM2).

6.2 OGCMs

Similar to AGCMs, OGCMs are based on the equation of fluid motion, conservation of salt and other substances, and the laws of thermodynamics. Also, these equations are expressed as a finite number of algebraic equations. Moreover, continuous functions are represented by their values at a discrete number of points, which brings up the discretization problem.

6.2.1 *Discretization of the equations of oceanic motions*

In a large class of OGCMs, the surface was assumed to be at a constant level where the vertical velocity vanished [Bryan, 1969]. The motivation for this "rigid-lid approximation" was to filter a mode that propagates horizontally with very high speeds hence requiring small time steps to avoid linear computational instability in the numerical solutions. However, the approximation required a global calculation that is inefficient in massively parallel computer architectures. Contemporary OGCMs use a "free surface" approximation, in which the height is computed by taken into account ocean dynamics and the mass transfer of water with the atmosphere. At material boundaries, such as the ocean bottom, the normal velocity is set to zero.

Vertical discretization of OGCMs is a very active area of research in which great progress has been made during the last two decades. Here, we will outline the fundamentals for a better appreciation of how the model produces the surface conditions required by the atmospheric model that completes the coupling system. An OGCM coupled to a simple atmosphere is expected to provide an SST field dynamically consistent with the surface heat fluxes it receives.

For the vertical discretization, most OGCMs use either the z coordinate, the σ coordinate (which divides the vertical distance between the ocean surface and bottom into discrete intervals), or the isopycnal coordinate based on the density [Griffies et al., 2010; see Fig. 6.1] or a combination of these systems at different depths. Using the z coordinate divides the ocean domain into boxes whose bottoms are located at fixed depths.

A terrain-following coordinate in an OGCM may introduce discretization errors in the horizontal pressure gradient terms as in AGCMs, and define very thin layers in shallow regions. An isopycnal coordinate results in increased resolution around the pycnocline, but in coarse resolution in regions with strong vertical mixing.

In the horizontal, many OGCMs distribute the variables on the Arakawa B-grid [Arakawa and Lamb, 1977]. In this grid, tracer quantities (potential temperature, salinity, and other tracers) are defined in the horizontal at different locations than the velocity components u and v. In contrast with the Arakawa C-grid [Arakawa and Lamb, 1977; see Fig. 5.1), both velocity components on the B-grid (see Fig. 6.2) are defined at the same location facilitating the calculation of the crucial Coriolis force term in the momentum equations. This is important in OGCMs, which can have grids with a size similar to the length scale of important phenomena such as the mesoscale ocean eddies. However, as resolution increases

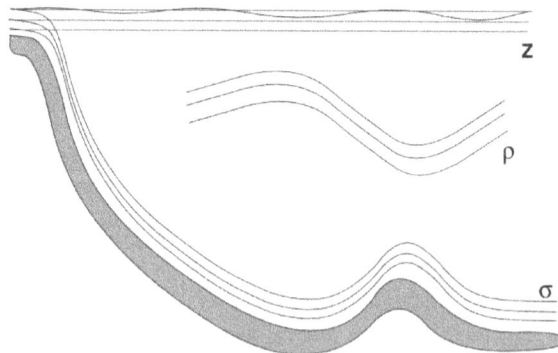

Figure 6.1.   Schematic of vertical coordinate systems used in OGCMs. (Figure 4 in Griffies et al. [2000].)

(B)

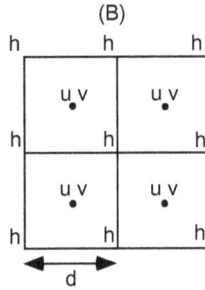

Figure 6.2.   Distribution of OGCM variables in the Arakawa B–grid. (From Fig. 4 in Mechoso and Arakawa [2015].)

have been made increasingly possible due to advances in computer technology, many OGCMs have switched to the C-grid.

The Bryan-Cox model [Bryan, 1969; Cox, 1984] used the B-grid in the horizontal. In the late 1980s, GFDL starting production of a series of OGCMs that have been labeled Modular Ocean Models (MOMs). In MOM1 the upper boundary condition is a horizontal rigid surface and two major components can be recognized: (1) OGCM/Baroclinic, which determines the deviations from the vertically averaged velocity, temperature and salinity fields, and (2) OGCM/ Barotropic, which determines the vertically averaged distributions of those fields (see Ch. 7). MOM2 has an option to relax the rigid lid approximation and solve the free surface equation. MOM3 updates several aspects of the physics and includes an implicit free surface. MOM4 includes a choice between quasi-horizontal coordinates and terrain following coordinates. MOM5 provides a C-grid option as well as a dynamically interacting Lagrangian sub-model. The most recent version, MOM6, uses the C-grid in the horizontal and offers several choices for the vertical coordinate, including the isopycnal coordinate as well as generalized vertical coordinates using a vertical Lagrangian remapping method [Bleck, 2002; Adcroft and Hallberg, 2006; Griffies et al., 2020]. Another descendant of the Bryan-Cox model [Bryan, 1969; Cox, 1984] is the Parallel Ocean Program (POP) development at the Los Alamos National Laboratory (LANL) originally for parallel computer architectures [Smith et al., 1992, 2010]. OGCM/POP became the ocean component in the community coupled models sponsored by the US NSF National Center for Atmospheric Research (NCAR). Further information on OGCM/POP is in http://www.cesm.ucar.edu/models/ ccsm2.0/pop/. The newest version of the NCAR Earth System Model (CESM) will incorporate MOM6 as the ocean component, with development ongoing at this time. The Hybrid Coordinate Ocean Model (HYCOM) is based on a vertical coordinate system that is isopycnal in the open, stratified ocean, and which smoothly transits to a terrain-following coordinate in shallow coastal regions, and to z-level coordinates in the mixed layer and/or unstratified seas.

This model makes use of the vertical Lagrangian remapping also used in MOM6 and pioneered for ocean modeling by Bleck (2002). Further information on HYCOM is in www.hycom.org.

Another example of a widely used OGCM is the Nucleus for European Modelling of the Ocean (NEMO). NEMO is a state-of-the-art modeling framework developed by a European consortium for research activities and forecasting services in ocean and climate sciences. The ocean component of NEMO is a primitive equation model adapted to regional and global domains. Prognostic variables are the three-dimensional velocity field, sea surface height, conservative temperature and absolute salinity. In the horizontal direction, the model uses a curvilinear orthogonal grid and in the vertical direction, a full or partial step z-coordinate, or σ-coordinate, or a mixture of the two. Various physical choices are available to describe ocean physics, including TKE, and GLS vertical physics. Within NEMO, the ocean is interfaced with a sea-ice model (SI3: Sea-Ice Integrated Initiative), and passive tracer and biogeochemical models (TOP). NEMO, SI, and TOP are documented in Madec and NEMO system team [2019], NEMO sea-ice working group [2018], and NEMO TOP working group [2019], respectively.

Reviews on various aspects of OGCMs include Griffies and Treguier [2013] (doi: 10.1016/B978-0-12-391851-2.00020-9), and Fox-Kemper et al. [2019] (doi: 10.3389/fmars.2019.00065)

6.2.2 *The sea surface temperature equation*

The prognostic equation for the temperature of the upper grid box in an OGCM is (see Fig. 6.3)

$$\rho c_p h \frac{\partial T}{\partial t} = F_{turb} + F_{rad} + F_{oceanL} + F_{oceanB} \tag{6.1}$$

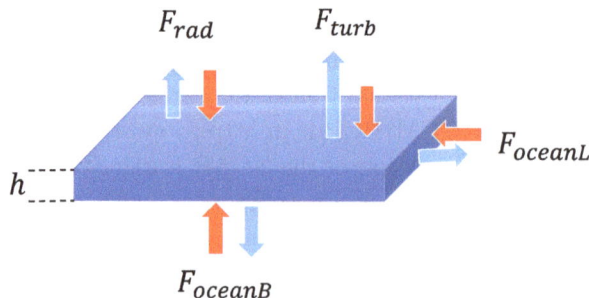

Figure 6.3.   Schematic of energy fluxes on the upper grid box of an ocean model.

or in more detail,

$$\rho c_p h \frac{\partial T}{\partial t} = F_{turb} + F_{rad} - \rho c_p h(\mathbf{v} \cdot \nabla T) - \rho c_p h \left(w \frac{\partial T}{\partial z} \right) + \dot{Q} \tag{6.2}$$

where $\rho = 1{,}025\,\text{km}^{-3}$ is a common choice for the reference density of seawater used in the Boussinesq approximation, $c_p = 3994\,\text{J} \cdot \text{kg}^{-1} \cdot \text{K}^{-1}$ is the specific heat of seawater at constant pressure, h is depth of the upper grid box, and T is SST. Further, F_{turb} is surface turbulent (sensible plus latent) heat flux, $F_{rad} = F_{cloud} + F_{clear}$, where F_{cloud} is net (downwelling minus upwelling) surface cloud-induced radiative flux, F_{clear} is net surface clear-sky radiative flux, and $\mathbf{v}$ is horizontal velocity in the upper layer.

In addition,

$$F_{lateral} = -\rho c_p h(\mathbf{v} \cdot \nabla T) \tag{6.3}$$

$$F_{bottom} = -\rho c_p h \left(w \frac{\partial T}{\partial z} \right) \tag{6.4}$$

are the lateral and bottom fluxes. Also, the last term in the right-hand side of Eq. (6.2) represents the contribution from the unresolved scales by the model's grid.

When running an OGCM by itself the surface forcing has to be provided. The estimation of turbulent fluxes at the surface F_{turb} was addressed in Ch. 5, where the aerodynamic formulations were presented. In terms of observational datasets, F_{clear} can be obtained from the satellite-based Clouds and the Earth's Radiant Energy System (CERES) Energy Balanced and Filled (EBAF) dataset version 4 [Kato et al., 2018], widely considered to produce the most state-of-the-art gridded observational estimates of surface radiative fluxes. F_{cloud} is the cloud radiative effect (CRE), which is defined as net (downwelling minus upwelling) all-sky minus clear-sky radiation at the ocean surface. Other possible sources of data for F_{turb} are the Objectively Analyzed Air-sea Fluxes (OAFlux) Project [Yu et al., 2008], the European Centre for Medium-Range Weather Forecasts (ECMWF) Interim Re-Analysis (ERA-Interim; Dee et al., 2011), the Climate Forecast System Reanalysis (CFSR) [Saha et al., 2010], and the Modern-Era Retrospective Analysis for Research and Applications, version 2 (MERRA-2) [Gelaro et al., 2017]. When coupled to an AGCM, this provides surface flux of momentum (wind stress), T (heat), S (water), and other material properties, while fluxes at the ocean bottom are neglected.

At this point, we briefly pause to emphasize that the thickness of the upper ocean layer h in Eq. (6.1) is not equivalent to the depth of the thermocline, as in simple ocean models such as an SOM (see Eq. 5.48). The concept of thermocline refers to an ocean layer in which the vertical thermal gradient is maximum. The depth of such a layer can certainly be computed in OGCMs such as MOM and POP, that use fixed vertical grids. Estimates of thermocline depths using observational data suggest that thermocline location in equatorial oceans can be represented by the 20°C isotherm. In OGCMs, however, the 20°C isotherm has been reported to be deeper and flatter than in observations [Castano-Tierno et al., 2018]. As such, it may have a reduced sensitivity to processes associated with atmosphere-ocean interaction and may not be representative of how well the model captures these processes. There is a conceptual parallel between these arguments and those used to contrast the PBL parameterization in AGCMs that consider the depth of this layer as either a diagnostic or a prognostic variable (see Ch. 5). HYCOM follows an intermediate approach. In this model, and starting at the surface, the program looks for the first "stairstep" in the density profile and defines the interface where this happens as the mixed layer base, with the turbulence parameterization having different formulations above and below the stairstep [Bleck, 2002]. A deeper look into these modeling aspects are beyond the scope of the present book.

6.2.3 *Incorporation of a sea-ice model to the OGCM*

Contemporary OGCMs include an active sea-ice model. This is generally fully coupled with the atmosphere above and the ocean below through exchanges of mass, heat, and momentum. The ice can have multiple categories and include several layers as well as a snow layer on top [Hunke, 2014]. The ice dynamics can be treated following visco-plastic rheology. A calculation of sea-ice growth rates is generally included. A prognostic sea-ice model with such capabilities allows sea-ice cover to respond to changes in the atmosphere and ocean states for long-range predictions. A Lagrangian approach to sea-ice model aimed at capturing sea-ice drift and deformation together with ice thickness and extent has been tested [Rampal et al., 2016].

6.2.4 *Wind stress on the ocean surface*

A key parameter in the calculation of surface wind stress over the ocean is the roughness length z_0, which depends on the state of the ocean surface where waves develop. A parameterization of the roughness length for strong winds was proposed by Charnock [1955] on the basis of the following nondimensional relationship

$$\alpha_{ch} = \frac{z_0 g}{u_*^2} \tag{6.5}$$

In this relationship α_{ch} is referred to as the Charnock parameter and was initially given a constant value, typically 0.018.

A more recent example of z_0 parameterization used at ECMWF is in Hersbach [2010]:

$$z_0 = \alpha_M \frac{\nu}{u_*} + \alpha_{ch} \frac{u_*^2}{g} \tag{6.6}$$

where ν is kinematic viscosity $(1.5 \times 10^{-5} \, \text{m}^2 \, \text{s}^{-1})$ and $\alpha_M = 0.11$. The second term in the right-hand-side of Eq. (6.6), which is Charnock's relationship in Eq. (6.5), dominates in conditions with sufficiently strong winds. The second term dominates in conditions with sufficiently light winds, and the contributions of both terms are comparable in cases with moderate winds. For neutral surface layers (See Ch. 5, Sec. 5.5.1), the wind velocity and drag coefficient are given by Eq. (5.20) and Eq. (5.25), respectively. If the contribution of the terms with the Monin–Obukhov length are neglected, then one can write,

$$\Phi_m = b_n = \log(1 + z/z_0); \quad \bar{u}_n = \frac{u_*}{\kappa} b_n; \quad C_D = \left(\frac{\kappa}{b_n}\right)^2 \tag{6.7}$$

where $\bar{u}_n$ is the equivalent neutral wind. In modeling applications, $\bar{u}_n$ can be taken as the wind at the lowest model level. In this case, the solution of (6.6)–(6.7) requires an iteration. However, it can be shown that iterations can be avoided by using approximate relationships of the form:

$$z_0^{fit} = \frac{z}{e^{b_n^{fit}} - 1}, \quad C_D^{nfit} = \left(\frac{\kappa}{b_n^{fit}}\right)^2 \tag{6.8}$$

where b_n^{fit} is an analytic function of the neutral wind speed at level height, and the Charnock parameter α_{ch} (see Hersbach, 2010 for details).

Linking the wind stress at the ocean surface with ocean surface waves is of great interest in modeling the coupled atmosphere ocean systems and predictions of the state of the sea for many applications such as marine weather forecasts (see Reichl et al., 2014). Ocean surface waves, and in particular wave breaking, modulate exchanges of momentum, heat, and mass between the atmosphere and the oceans. An extensive discussion on the interactions of ocean winds and waves is in Janssen (2004). The following paragraph presents two examples of these links.

Taylor and Yelland (2001) used datasets representing sea-state conditions ranging from strongly forced to shoaling to propose an expression based on wave steepness,

$$\frac{z_0}{H_s} = 1200 \left(\frac{H_s}{L_p}\right)^{4.5} \tag{6.9}$$

where H_s is the significant wave height and L_p is the wavelength at the peak of the wave spectrum. of the dominant waves. Drennan et al. (2003) on the basis of data from field experiments representing a variety of conditions, proposed the following relationship,

$$\frac{z_0}{H_s} = 3.35 \left(\frac{u_*}{c_p}\right)^{3.4} \tag{6.10}$$

where c_p is the wave phase speed at the peak of the spectrum and c_p/u_* expresses wave age. Wave-age scaling predicts stronger changes in roughness associated with fetch or duration than steepness scaling.

Current models of the coupled atmosphere-ocean model include a module for surface wave prediction. The Community Earth System Model version 2 (CESM2; in Danabasoglu et al., 2020) incorporates a version of the NOAA WaveWatch-III ocean surface wave prediction model. The waves allow for including Langmuir, or wave-driven, turbulence in the ocean model, which has several benefits. The ECMWF's Integrated Forecasting System (IFS) also includes a surface wave model.

6.3   Examples of OGCMs coupled to simple models

This section compiles selected examples of applications of HCM2s to research on the ocean variability in the Pacific and Atlantic basins. High-frequency atmospheric noise is assumed not to play an essential role in the scientific problems addressed with such models. In each case, we highlight the especial formulations required to produce the surface forcing fields and implement the coupling.

6.3.1   *Tropical Atlantic OGCM coupled to an empirical atmosphere*

The first example is an HCM2 applied by Chang et al. [2001] to address the following question: How much of the SST variability in the tropical Atlantic can be explained as a direct forced oceanic response to internal atmospheric variability with a secondary contribution from local air-sea interactions? It is widely accepted that atmosphere-ocean feedbacks play key roles in the tropical Pacific variability. Strong air-sea coupling can also be expected in the tropical Atlantic in association with the feedbacks introduced in Ch. 2, particularly the wind-evaporation-SST (WES) feedback. The question posed, therefore, addresses the possibility of fundamental differences between two tropical oceans.

The ocean component in this case is MOM3 with a domain set to the Atlantic basin between 30°S and 45°N. A relaxation term towards a prescribed annual mean SST distribution (T_{mean}) from the observation is included by adding to the surface heat flux a Newtonian damping term $-\gamma(T-T_{mean})$, where γ is a constant representing an inverse relaxation time scale to climate state.

This ocean model is coupled to an empirical atmosphere based on the covariant patterns of anomalies in SST, surface wind stress, and heat flux (see Ch. 4). It is assumed that the spatial pattern of SST anomalies produced by the OGCM at time t determines those of the other anomalies. This special interdependence is implemented through application of a procedure called singular value decomposition [SVD; Bretherton et al., 1992; An and Wang, 2000; Syu et al., 1995]. Starting from the time series of spatial SST anomalies patterns provided by an observational dataset, SVD produces a number of modes e_n (Eq. (4.60a)) and the corresponding ones for the other anomaly fields such that pairs of modes explain as much as possible of the mean squared temporal covariance between fields. Once this preliminary work is completed with the observational datasets, the OGCM runs producing a pattern of SST anomalies for each time t, which is projected onto the modes e_n resulting in coefficients α_n. The coefficients for the wind and heat flux anomaly fields are also set to α_n, except for normalization factors. All forcing fields can, therefore be reconstructed from the respective coefficients and used to force the OGCM to march in time.

One important task to accomplish before running the model is to set up the specifics of the background forcing for an OGCM run that provides the reference for the anomalies. In Chang et al. [2001] the background forcing corresponds to the monthly averaged surface stresses and heat flux derived from the output of a long AGCM run with a prescribed distribution of SST from an observed climatology. Lastly, parameters to be explored have to be chosen. Two scaling dynamic and thermodynamic coupling parameters are introduced as factors: (1) α for the wind stress, affecting the strength of dynamic feedbacks, and (2) β for the surface heat flux, affecting the strength of thermodynamic feedbacks.

A parameter sweep of experiments performed with the model showed that oscillations with different periods could be obtained in the Atlantic for certain combinations of the parameters α and β, particularly if the latter reaches a threshold value. This suggests that in a strong thermodynamic coupling regime local air-sea feedbacks in the tropical Atlantic can support a self-sustained decadal oscillation resembling the observed Tropical Atlantic Variability [TAV; see Yu et al., 2020]. Adding to the resemblance, the oscillation exhibits strong cross-equatorial SST gradients and meridional wind variability. The role of positive feedbacks involving surface heat flux and SST and of negative feedbacks involving ocean currents, as well as the regions where the competition among positive and negative feedbacks is crucial were highlighted. Moreover, the time evolution of the SST anomalies could be produced and compared with the observation. The results also suggested that local air-sea coupling in the tropical Atlantic may not be strong enough to maintain a self-sustained oscillation, and additional perturbations from the extratropics might be required.

The findings that were described illustrate the usefulness of hybrid models as tools that permit the performance of ensemble numerical experiments that provide insight into the fundamentals of climate events, and yet do require substantially less computer resources than those of an OGCM coupled to a full AGCM. The previous discussion, however, also illustrates the multiple caveats that must be kept in mind for a meaningful use of this tool, which is not so simple after all. For example, dynamical processes in the atmosphere that are independent of SST forcing are largely filtered out. Moreover, an analysis of the type described could only address modes of variability that could be represented by the SVD selected for expansion. It is possible that interannual modes [e.g., Atlantic Niño; Zebiak, 1993] are missed for this reason. In general, modeling approaches that strongly depend on observational data for constraining the simulations may suffer from the relatively short observational record, which may not suffice to fully capture the signal under investigation.

6.3.2 OGCM coupled to a two-level atmosphere for ENSO studies

Another HCM2 for ENSO studies has been used by Neelin [1990]. In this case the OGCM component is an adaptation of a version of the NOAA/GFDL MOM1 used by Philander et al. [1987] for analyses of tropical Pacific variability. The adaptation keeps the OGCM domain restricted to the Pacific Ocean between 30°S to 50°N and continued ignoring salinity effects, but horizontal resolution is reduced by a factor of about three, there are 10 levels in the vertical, and no continental outlines are included. The atmospheric temperature just above the ocean surface is prescribed from an observational dataset. In preparation for the coupled run, a climatology of the OGCM was obtained by running the model using climatological wind stress from the observation and radiative flux at the ocean surface given as an idealized function of latitude. Outside the ocean model domain, the SSTs are set to the zonal average of the climatological values over the basin.

The atmospheric component of this hybrid coupled model is a steady-state two-level model. See schematic of vertical structure in Fig. 6.1. For use in the coupled mode, the calculation of surface sensible heat flux assumes that air temperature is specified from an observed climatology, i.e., air temperatures do not respond to SST anomalies. The low level moisture (q_2) is parameterized as a function of SST in the following way:

$$q_2 = \alpha q_{sat}(SST - \delta T), \tag{6.11}$$

in which α is a constant (i.e., 0.8), q_{sat} is saturation mixing ratio at 1000 hPa, and a constant value of δT represents the difference between the temperature of the ocean surface and the overlying air. The net radiative flux at the top of the atmosphere and ocean surface are specified from the results of an AGCM simulation (with prescribed SSTs). Otherwise, radiative fluxes are parameterized as a Newtonian damping to a prescribed reference state.

The assumptions described in the previous paragraphs allow for the calculation of the difference $(\bar{F}_B - \bar{F}_T)$ between the time mean net energy fluxes at the bottom and top of the atmosphere. Starting from here, the low level convergence can be estimated according to the following expression [Neelin and Held, 1987],

$$-\nabla \cdot \mathbf{V}_2 = \frac{\bar{F}_B - \bar{F}_T}{\Delta s - q_2} \qquad (6.12)$$

where Δs is to be determined from comparison of model integrations and observation and q_2 given by Eq. (6.11). Moreover, a simple parameterization for the surface stress is used [Neelin, 1988],

$$\tau_s = K_m \mathbf{V}_2 \qquad (6.13)$$

where K_m is an ad hoc parameter with dimensions of inverse time.

Coupling is asynchronous (see Ch. 8) and carried out with a coupling interval of 15 days. At a coupling instance, oceanic conditions averaged over the previous period are passed to the atmospheric model which returns the steady-state atmospheric response to these conditions. A one-way flux correction is applied according to which the atmospheric model climatology is subtracted from the atmospheric model response and replaced by observed climatological wind stresses corresponding to annual average conditions.

In experiments with this model using different parameters and initial conditions, the wind stress that is passed by the atmospheric model to the ocean model was further altered by multiplication by a constant factor, which is labeled "relative coupling coefficient" and varies in the 0.6–1.0 range. Thus, the wind stress to the ocean can be weakened or strengthened for the purposes of conceptual understanding. Figure 6.4 shows a complex SST evolution in which a quasi-quadrennial oscillation appears superimposed to coupled Kelvin waves with a period near six months that weaken in time. More runs with other parameters showing similarly complex evolutions are presented in Neelin [1990].

6.3.3 *Reduced gravity ocean model coupled to a statistical atmosphere*

A coupled model for the Pacific basin that incorporates several additional physical processes to the ones mentioned so far was developed by Zhang [2015]. In this case the ocean component is based on the primitive equation model with the reduced gravity approximation developed by Gent and Cane [1989]

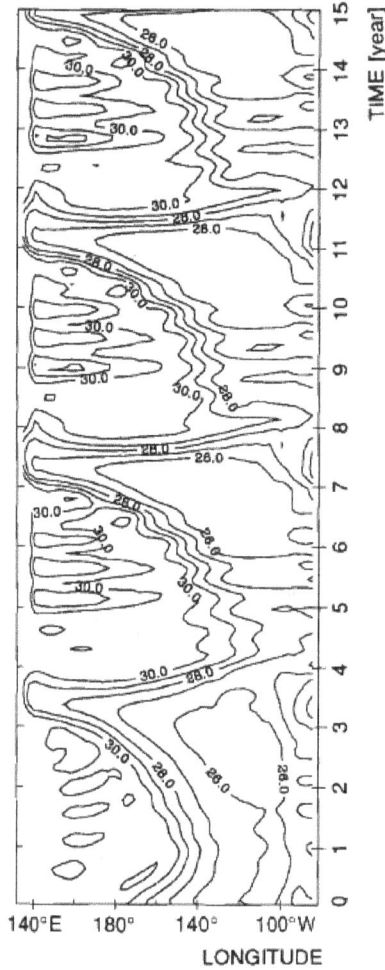

Figure 6.4. SST (°C) along the equatorial Pacific produced by an HCM2 consisting of a tropical Pacific version of the NOAA/GFDL MOM1 coupled to a steady-state two-level model. (Figure 15 in Neelin et al. [1992].)

in a configuration that covers the Pacific from 40°S to 60°N and from 124°E to 76°W. The complexity of this model justifies including this example in the present subsection although the ocean component is not exactly an OGCM. The atmospheric component is similar to the one described in Sec. 6.3.1.

A schematic diagram illustrating this model is shown in Fig. 6.5. A stochastic contribution (τ_{SF}) is added to the surface wind stress. The effects of surface freshwater flux (FWF), obtained from the difference between precipitation and evaporation (P–E), are included in the expressions for sea surface salinity (SSS) and buoyancy flux (Q_B). Processes internal to the ocean are also considered, such as those associated with Tropical Instability Waves [τ_{IW}; Zhang and

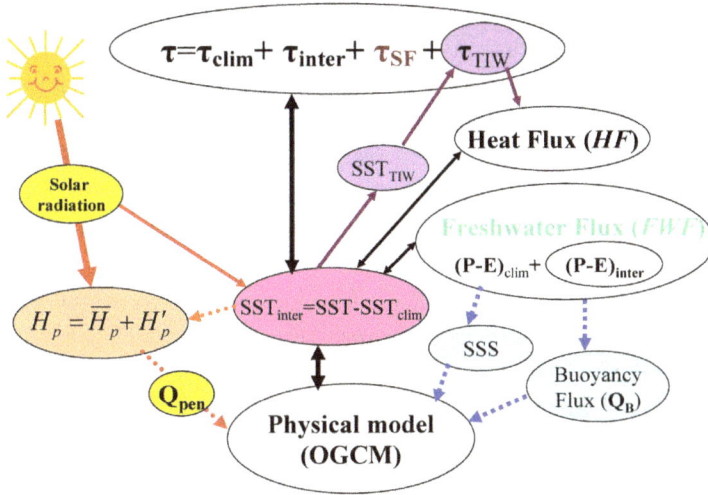

Figure 6.5. Schematic diagram illustrating the hybrid coupled model consisting of an OGCM and a simplified representation of the atmosphere in Zhang (2015). Symbols are explained in the text. Fields are separated into their climatological part and interannual anomaly part. (Figure 1 in Zhang [2015].)

Busalacchi, 2009b] and other processes (H_p) that influence the penetration of solar radiation into the upper ocean (Q_{pen}).

This model is reported to produce realistic simulations of the atmosphere-ocean system in the Pacific with an efficient utilization of computer resources. Despite the addition of a stochastic component to the ocean forcing, however, the system variability seems to be limited meaning that other relevant processes may be missing. For example, the simulated ENSO events are far too regular, dominated by two-year oscillation. Recent work has demonstrated that interannual variability in the tropical Atlantic can influence that in the Pacific [Rodriguez-Fonseca et al., 2009]. Such remote effects are not included in basin-type configurations.

6.3.4 *Primitive equation ocean model coupled to an empirical atmosphere for ENSO studies*

A fully nonlinear primitive equation ocean model coupled to a statistical atmosphere developed by Barnett et al. [1993] was introduced in Sec. 4.7 in this book.

6.3.5 *OGCM coupled to a statistical atmosphere for AMOC studies*

An OGCM coupled to a statistical atmosphere model has been used to study the effects of ocean-atmosphere feedback on the stability of the Atlantic meridional overturning circulation [AMOC; Cimatoribus et al., 2012a,b]. Strong atmosphere-ocean feedbacks may remove the multiple equilibria regime by overcoming the effects of the salt-advection feedback, which is a major contributor in the AMOC's hysteresis behavior to external freshwater forcing in uncoupled models and intermediate complexity coupled models. For more information on the possibility of AMOC collapse the reader is referred to Cazenave et al. [2021] and references therein. Unlike the other examples in this subsection, the time scales involved are much longer than those of ENSO.

Designing a statistical atmosphere for climate processes at very long time scales poses additional challenges because the observational datasets available to find the significant correlations among variables needed to build a statistical atmospheric model are not much longer than a few decades. Obtaining data from fully coupled atmosphere-ocean GCMs (CGCMs; see Ch. 7) is not feasible at the present time because multi-thousand year runs are required. Cimatoribus et al. [2012a,b] go around these problems by using an intermediate complexity HCM in which the coupling allows for feedbacks from processes specifically at work for Atlantic meridional overturning variability in a fully coupled CGCM. In this case, such feedbacks came from the output of EMIC SPEEDO [Severijns and Hazeleger, 2010], an intermediate complexity coupled atmosphere/land/ocean/sea-ice GCM that captures an abrupt collapse of the AMOC. The OGCM component of EMIC is CLIO [Goosse and Fichefet, 1999], which includes the LIM sea-ice model (Graham and Barnett, 1987). The atmospheric component is a modified version of the Speedy AGCM [Molteni, 2003]. Ocean-atmosphere feedbacks to keep statistical equilibrium state of AMOC were extracted from a steady state run of SPEEDO, while feedbacks for the AMOC collapse state came from flash-water hosing run of SPEEDO that produce the collapse.

Two statistical models of the atmosphere could then be built using the linear regression method and the two datasets. Let ϕ be a surface flux field (heat, freshwater, and momentum). In the linear regression model, this flux field is written as

$$\phi(i,j) - \overline{\phi(i,j)} = p(i,j)(SST(i,j) - \overline{SST(i,j)}) \qquad (6.14)$$

where p is a parameter to be fitted by regression; i and j are horizontal grid indices; and overbars indicate time average. The HCM2 resulting from coupling each statistical atmosphere to the OGCM captures the two different AMOC regimes (equilibrium state and collapse) using much less computational resources than the intermediate complexity model.

Cimatoribus et al. [2012b] use this HCM2 to investigate the sensitivity of the AMOC to freshwater anomalies in the South Atlantic. The authors suggest that nonlinear fitting rather than linear fitting does not add to the results, while the inclusion of atmospheric noise and lagged atmospheric response may significantly improve the representation of atmosphere-ocean interaction.

6.3.6 *AGCM coupled to a Slab Ocean Model in the tropical Atlantic and to an OGCM elsewhere*

The last example in this series is really a transition with the subject of the next chapter. The model here consists of an AGCM coupled to a Slab Ocean Model (SOM) in the tropical Atlantic and to an OGCM elsewhere; i.e. a combination of an HCM1 and of a full coupled GCM. This model was used by Zhang and Delworth [2006] to assess the impact of low frequency Atlantic variability on the variability in other ocean basins through the atmospheric bridge. The SOM, therefore, was forced with heat fluxes that contained the signature of the Atlantic Multidecadal Oscillation [AMO; Yu et al., 2020]. Otherwise, the Atlantic is unable to generate oceanic Kelvin waves propagating to other ocean basins and can only produce inter basin interactions via the atmosphere. The results of this study suggest that the multidecadal variability in the AMO can affect the characteristics and phase transitions of the Pacific Decadal Oscillation [PDO; e.g. Yang et al., 2020]. It can also contribute to multidecadal variations in the Indian Ocean basin.

6.4 Perspectives

OGCM development is a particularly active field. This is propelled by the key importance on resolving the ocean circulation in a very broad range of applications. The great variety of geographical domains of interest, such as the continental shelves and the polar regions, have motivated work on nesting and downscaling methods [e.g., Ringler et al., 2013; Danilov, 2013]. The importance of resolving mesoscale motions [Blein et al., 2020] has pushed the limits of using computers to provide simulations with a horizontal grid spacing in the order of a few kilometers. Moreover, the view that for long-range prediction of climate studies oceans are expected to provide the atmosphere with the SST field has been refined to the need to appropriately include the effects of sea-ice effects and surface waves.

The challenges mentioned in the previous paragraphs highlight the need for intermediate tools to address the conceptual understanding and provide guidance on how to proceed. In this sense, HCM2s have proved their importance as scientifically sound platforms to assess the sensitivity of the coupled atmosphere-ocean system to aspects that are either not well quantified or have not been investigated in depth. The examples reviewed in this subsection reveal a wide range of strategies that complement each other. The use of simple atmospheric models means that HCM2s are less demanding of computer resources than OGCMs coupled to AGCMs. There are several trade-offs to this merit: (1) wind stresses are replaced with an algorithm that uses an observed climatology, (2) surface temperature is specified from the observations, and (3) the net radiative flux at the top of the atmosphere is prescribed. Moreover, the coupling fields are through multiplication by a relative coupling coefficient. In addition, depending on configuration and computer architectures, it is possible that full AGCMs require much less resources than a full OGCM thus reducing the appeal of HCM2. Therefore, it can be reasonably anticipated that work with HCM2s will continue to be encouraged in teaching and research institutions.

Chapter 7

Atmosphere-Ocean Coupled General Circulation Models

7.1 Introduction

In general circulation models of the coupled atmosphere-ocean system (CGCMs) both the atmospheric and oceanic components are GCMs (AGCMs and OGCMs, respectively). AGCMs were introduced in Ch. 5 and OGCMs in Ch. 6. CGCMs represent the top complexity in terms of numerical tools for synthesis of a number of key interacting processes within the global climate system. The presentation in this chapter is really a snapshot of a huge research field as models evolve continually because representations of physical processes as well as numerical schemes are actively revised and new ones are incorporated. Moreover, not all processes are covered in this chapter, notably the effects of the aerosol.

CGCMs integrations follow the initial-value problem approach. Therefore, they can be used both for climate simulation and numerical weather prediction (NWP). The dependence of the solution on initial conditions is less emphasized in the former applications, while it is crucially important in the latter applications [Mechoso and Arakawa, 2015]. The following sections dedicate particular attention to the fields exchanged between CGCM components to realize the coupling.

7.2 Fundamentals of AGCM and OGCM coupling

Coupling of AGCM and OGCM codes is achieved through data exchanges at the atmosphere-ocean interface and sets of instructions on synchronization. In this section we look at the basic case, in which the AGCM passes surface wind stress, heat and water fluxes to the OGCM, and the latter passes sea surface temperature (SST) to the former model. Even at this level of simplicity, the implementation of coupling poses an important number of scientific/technical questions.

If the AGCM and OGCM are both running in a single node of a computer, as it was the case in pioneering applications (see Ch. 8), then the models would be run sequentially. In this scenario, the AGCM could be assigned the "master" role for managing calls to regridding routines and directing the data flow at the instances of exchange established by coupling intervals, and the OGCM would become a subroutine of the AGCM. In such an arrangement, model codes would execute sequentially and exchange information corresponding to the air-sea interface (i.e., surface fluxes from the AGCM and SST from the OGCM).

When the possibility of running the AGCM and OGCM codes in separate computer nodes became available, coupling software was especially designed to allow for data exchanges between subcomponents in order to exploit parallelism of execution. Chapters 8 and 9 give detailed presentations of the general coupling problem together with technologies and the special software currently used by several modeling groups. As an introduction to these chapters, we examine the case in which only two computer nodes are available.

The availabilty of two computer nodes introduces an additional feature of parallelism because — conceptually — two components can be recognized in AGCMs: "Dynamics" and "Physics". Physics produces the drivers for the atmospheric motions through "parameterizations" (see Ch. 5), and hence is computed first. Calculations in this component are generally performed on a latitude-longitude-height grid. The major physical parameterizations (i.e., radiation and convection) currently operate on separate vertical columns of the atmosphere. Very little communication is required between processors if a two-dimensional grid partition in the horizontal (longitude-latitude) is used and, therefore, the code is almost "embarrassingly" parallel. The drivers of the flow provided by Physics are transferred to Dynamics, which computes the evolution of the atmosphere based on the equations that govern fluid motion. The special role of each model component allows for task parallelization. AGCM/Dynamics can run in parallel with the OGCM provided that the SST is fixed to the last field available from this component. Figure 7.1 is a schematic of this running strategy, under the assumption that overheads due to distribution of the calculation can be minimized and that model components running in parallel can be approximately balanced.

We note that this approach requires some knowledge of the AGCM's code in order to separate the Dynamics and Physics parts. Also, the possibility of such a separation is not necessarily universal. For these reasons, coupling strategies used at present by many modeling groups deal with the full AGCM and OGCM as individual tasks, and focus on the knowledge of the fields produced by the models rather than on their internal structures.

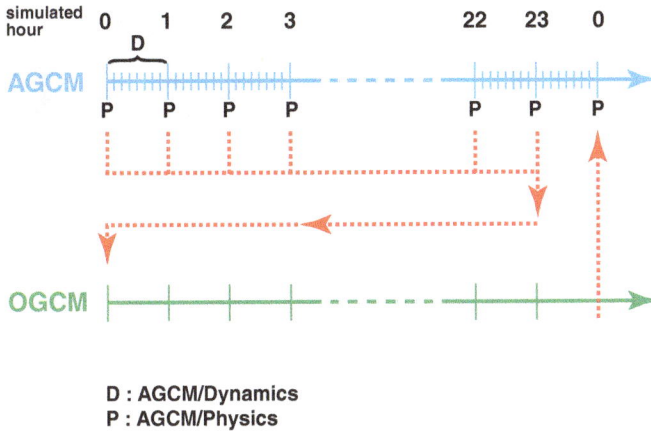

D : AGCM/Dynamics
P : AGCM/Physics

Figure 7.1.   Diagram showing the procedure of coupling an AGCM in one computer node to an OGCM in another computer node during a CGCM run with time increasing from left to right. The short tick marks in the horizontal line labeled AGCM represents a calculation of the Dynamics component ("D"), while long tick marks correspond to calculations of the Physics component ("P"). In this example, the Dynamics is computed 8 times between calls to Physics, and the Physics time step is 1 hour. The tick marks in the horizontal line labeled OGCM represents a calculation by this model with a time step of 1 hour. The dotted arrows between the AGCM and OGCM lines point to the instances of data exchanges for a coupling interval is 1 day (24 hours), (Figure 2 in Mechoso et al. [1993], © American Meteorological Society, used with permission.)

7.3   Current issues on CGCM performance

A wide variety of AGCMs and OGCMs have been coupled so far, varying in numerical algorithms and parameterizations as well as in horizontal and vertical resolutions. Increased computer power has allowed operation at high and very high resolutions up to fractions of kilometers. However, extensive numerical experimentation with state-of-the-art CGCMs has shown that their simulation of the present climate shows important systematic errors (biases). These have been reported in several papers for a number of years and elimination of these errors has been and still is the motivation of much research [Mechoso et al., 1995; Davey et al., 2002; Wang et al., 2014; Zuidema et al., 2016].

A representation of the SST biases by CMIP5 models in reference to the period from 1900 to 2005 is shown in Fig. 7.2, which is taken from Wang et al. [2014]. The magnitude of these biases can be as large as several degrees Celsius. In the extratropics, SSTs are generally too low in the Northern Hemisphere and too high in the Southern Hemisphere. In the tropics, SSTs are typically too high in the southeastern Pacific and Atlantic and too low in the equatorial and tropical southwestern Pacific. In general, the patterns of these biases are

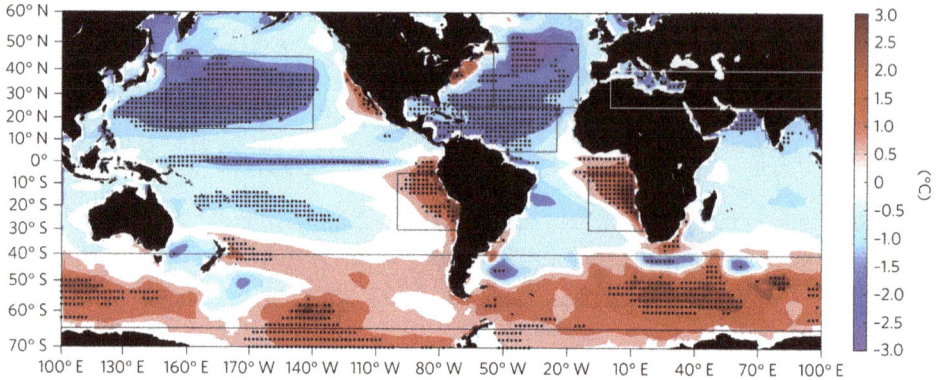

Figure 7.2.   Annual-mean SST bias averaged in 22 climate models. The locations where 82% of the models agree on the sign of the bias is highlighted by dots. (Figure 1 in Wang et al. [2011].)

largely independent of season, but amplitudes can vary. Along the equator, cold SSTs extend too far west from the continents. In the Atlantic, the sign of the simulated annual mean SST zonal gradient is opposite to the observations [Richter and Xie, 2008; Davey et al., 2002; Richter et al., 2011; Zuidema et al., 2016]. Cold SSTs in the northwest Atlantic and Caribbean Sea discourage the strong diabatic heating and ascent over these regions during boreal summer and the associated descent above the tropical southeastern Pacific in a Hadley-type, interhemispheric circulation [Lee et al., 2013]. That is, SST biases in the tropical North Atlantic can be remotely linked to those in the tropical southeastern Pacific. Sasaki et al. [2014] showed that the removal of the equatorial Atlantic SST bias leads to an improvement in the representation of the tropical Pacific climatology.

The simulation of tropical rainfall by CGCMs also has pervasive biases, which result in a too symmetric structure across the equator. This bias manifest itself as a double intertropical convergence zone (ITCZ) [Mechoso et al., 1995]. The double ITCZ bias remains a feature of concern in CGCMs [Bellucci et al., 2010; Li and Xie, 2014; Oueslati and Bellon, 2015; Xiang et al., 2017].

Many studies have addressed the causes for the tropical SST biases of CGCMs. Among the biases, the too high SSTs over the southeastern tropical Pacific and Atlantic have received considerable attention. The regional underrepresentation of low-level stratocumulus decks has been mentioned as one of the potential causes for this error through locally excessive shortwave radiative flux into the ocean [Ma et al., 1996; Huang et al., 2007; Hu et al., 2008; Toniazzo and Woolnough, 2014; Voldoire et al., 2014]. We return to this suggestion later in this chapter.

Other studies have addressed the westerly wind bias at the surface over the equatorial Pacific and Atlantic. Such a wind bias results in a simulated thermocline that is too shallow in the west and too deep in the east inhibiting the development of the equatorial cold tongue [Richter et al., 2011; Voldoire et al., 2014]. It has been suggested that the warm SST bias in the equatorial western Atlantic can be advected eastward along the equator and southward along the African coast through propagating oceanic downwelling Kelvin waves, contributing to further warming of the southeastern tropical Atlantic [Toniazzo and Woolnough, 2014; Goubanova et al., 2019]. The existence of the westerly wind bias and associated errors in wind stress in the simulations has been attributed to model difficulties with the PBL parameterization over the ocean [Zermeño-Diaz and Zhang, 2013] and the convective parameterization over the oceanic warm pools and tropical continents [Zermeño-Diaz and Zhang, 2013; Richter et al., 2012]. Too weak convection over the Pacific warm pool or tropical South America results in weaker oceanic easterlies to the east as well as in local reductions of upward motion and cloudiness, which reduce sinking motion over remote stratocumulus decks. Song and Zhang [2018] conclude that modifications in convection schemes can be a major contributor to the elimination of the double ITCZ bias. According to the consensus view, it may be meaningless to assign the double ITCZ bias to a single cause as several feedbacks are likely to contribute to it.

The global extent of GCM biases has also led to conjectures on the relationship among remote biases. Hwang and Frierson [2013] argued that errors in the Southern Ocean could be accountable for a large part of the double ITCZ bias. The erroneous interhemispheric energy fluxes by the ocean can shift the simulated raising branch of the Hadley cell and the ITCZ too far south [Kang et al., 2008; Li and Xie, 2014]. Mechoso et al. [2016] showed that this remote link is stronger in models with a high sensitivity to the stratocumulus-SST feedback (see Ch. 1). Similarly, the cold SST bias in high latitudes of the Northern Hemisphere can produce a cooling of the tropics through increased poleward eddy energy transport in the atmosphere. Wang et al. [2014] connected the extratropical SST biases worldwide to a too weak simulation of the Atlantic Meridional Overturning Circulation (AMOC).

CGCM biases can hinder the operation of key dynamical feedbacks. The models clearly struggle with the major components of the tropical interannual variability, vis-à-vis the Pacific and Atlantic Niños [Richter et al., 2014]. In addition, it has become increasingly apparent that SST anomalies in the tropical Atlantic can impact the tropical Pacific, which means that notable errors in the former region can impact the model's representation of climate variability in the latter region [Rodriguez-Fonseca et al., 2009]. The tropical biases can also mask the footprints of climate change in long simulations with perturbations in the composition of the atmosphere [McGregor et al., 2018; Kajtar et al., 2018].

7.4   CGCMs as laboratories for hypothesis-testing experimentation

GCMs have been widely used to perform numerical experiments aimed to test hypotheses on fundamental aspects of the climate system. The procedure is based on sensitivity studies in which the results of a "control" simulation are contrasted with those in which the model is artificially modified. An increased capability of computer resources allows for the performance of ensemble experiments and the assessment of statistical significance of the differences among results.

Early AGCM examples of this methodology are given by studies to assess the role of mountains on the general circulation of the atmosphere by Mintz and Arakawa [1965], Kasahara and Washington [1971], Manabe and Tepstra [1974] and Hahn and Manabe [1975]. In these papers, a control simulation by an AGCM is compared with a "twin" experiment in which selected mountains are removed from the boundary conditions; an experiment that cannot obviously be performed in the real world. A deep understanding of the physical problem under investigation and of the AGCM are required to design meaningful experiments that produce consequential results. For example, some early AGCMs included drastic assumptions on the treatment of cloudiness that may significantly obscure the findings.

7.4.1   *Testing hypothesis on the causes of the tropical SST biases*

To explore the sources of systematic errors in CGCMs the methodology consists of contrasting a control simulation (with the biases) with another simulation in which the biases are artificially corrected or weakened. Such an exercise, if carefully planned, can give very useful clues on the physical processes that are not properly captured by the model but does not directly address the model's intrinsic ability to reduce the errors.

Ma et al. [1996] examined the role of stratocumulus in the southeastern tropical Pacific climate using a version of the UCLA CGCM. A control simulation that underestimated regional stratocumulus was compared with another experiment in which a stratocumulus deck was prescribed to persistently cover an ocean region off the Peruvian coast. The results demonstrated that enhanced (and more realistic) cloud amounts can significantly reduce the SST bias over much of the eastern tropical Pacific south of the equator and even along the equator well into the central Pacific. The double ITCZ bias was also reduced, and surface winds along the eastern coast of South America became more realistic.

We next present the results of a sensitivity study using the UCLA AGCM with a focus on the stratocumulus decks along the Namibian coast in the

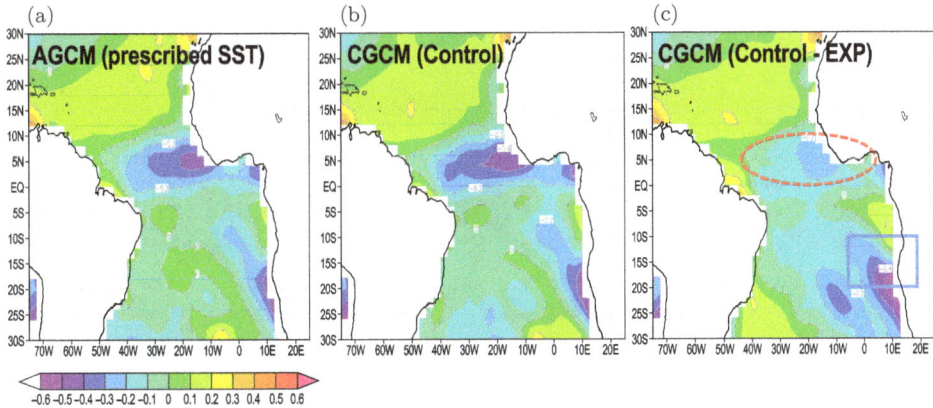

Figure 7.3. July mean meridional wind stress errors from a simulation with (a) the UCLA AGCM using prescribed SST corresponding to an observed climatology (AGCM), and (b) a long simulation with the UCLA CGCM (Control). Panel (c) shows the difference between Control and another simulation (EXP) using the same CGCM in which stratocumulus fraction and liquid water path approximate the values in the observation inside the blue rectangle. Units are dyn cm^{-2}. (From Xiao and Mechoso, personal communication.)

southeastern Atlantic [Xiao and Mechoso; personal communication]. In this case, the idealized experiment (EXP) prescribed the time-varying value of stratocumulus fraction and liquid water path based on the observation over the region between 10°S–20°S, and from 5°W eastward to the African coast.

Figure 7.3 shows the mean meridional wind stress errors for July from a long simulation by the atmospheric component of the model using prescribed SST (AGCM), a long CGCM simulation (Control), and the difference between Control and EXP. Mean flow at low levels along the Namibian coast is southerly in the observation and both AGCM and Control although the latter are too weak. The AGCM produces too weak winds along the Namibian coast and from Africa west to the north of the equator. Control reproduces the error patterns with slighter stronger intensities. EXP reduces these errors not only in the region where stratocumulus clouds are prescribed (blue box) but also north of the equator (dashed red oval). The errors in SST and precipitation over the tropical Atlantic are also reduced (not shown). These results provide further support to the notion that difficulties of CGCMs in the simulation of low-level marine clouds contribute to both local and remote model errors.

Another approach to investigation of the causes for CGCM biases is based on examining the time evolution of errors in experimental predictions, i.e., simulations initialized with observations (e.g., Huang et al., 2007; Toniazzo and Woolnough, 2014; Voldoire et al., 2014). Figure 7.4 presents experimental predictions of SST anomalies in a region of the equatorial Pacific known as Niño 3.4 (NINO3.4) by versions of the CGCM at the European Centre for Medium Range Weather Forecasts (ECMWF). The results represented by the

NINO3.4 mean absolute SST

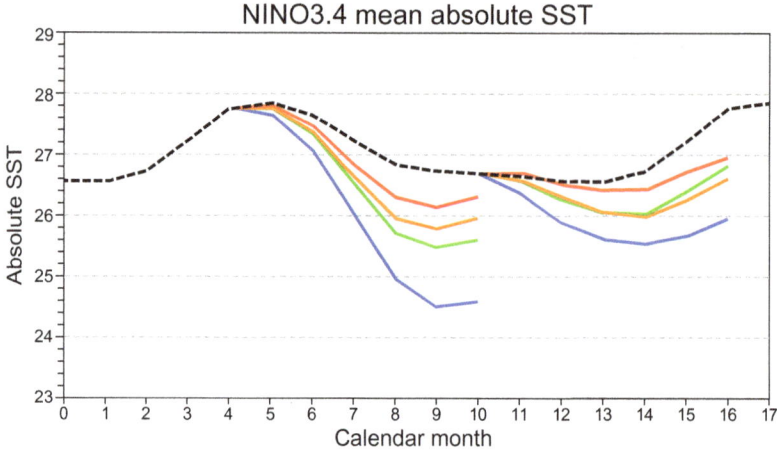

Figure 7.4. Experimental predictions of NINO3.4 SST starting in May and October by versions of the CGCM at ECMWF. Red curves: High resolution atmosphere and ocean. Orange curves: High resolution atmosphere and low resolution ocean. Green curves: Low resolution atmosphere and ocean. Blue curve: Previous version of the model with intermediate resolution. Black dashed line: Observation. (Figure 23 in Stockdale et al. [2018]. ECMWF Technical Memorandum No. 835.)

red curve are clearly superior to the others. Inspection of the difference between experiments suggested that about half the improvement comes from upgrading the model parameterizations, while the other half comes from the increasing resolution. A closer look at the results [Stockdale et al., 2018] has shown that the improvement coming from the increase in ocean resolution is related to a more realistic thermocline feedback (see Ch. 1). The increased resolution of the atmosphere also contributes, especially for forecasts initialized in May, which appears to be a consequence of a reduction of the zonal wind bias in the western Pacific.

The results outlined in the previous paragraph further confirm the expectation of better performance with increased resolution. In this regard, we mention that a seasonal timescale global simulation of the Earth's atmosphere with 1 km average grid spacing has been completed. The simulation was run with an adapted version of the ECMWF Integrated Forecasting System (IFS) on Oak Ridge's Summit computer — the fastest computer system in operation. Nevertheless, the need for future model improvement is an undisputable message.

7.4.2 Simulations that address enhancing prediction skill

Several papers using observations strongly suggest that interannual variability in the tropical Atlantic has significant impacts on the Pacific since the 1970s [e.g., Rodriguez-Fonseca et al., 2009]. In particular, Niña conditions in the Atlantic

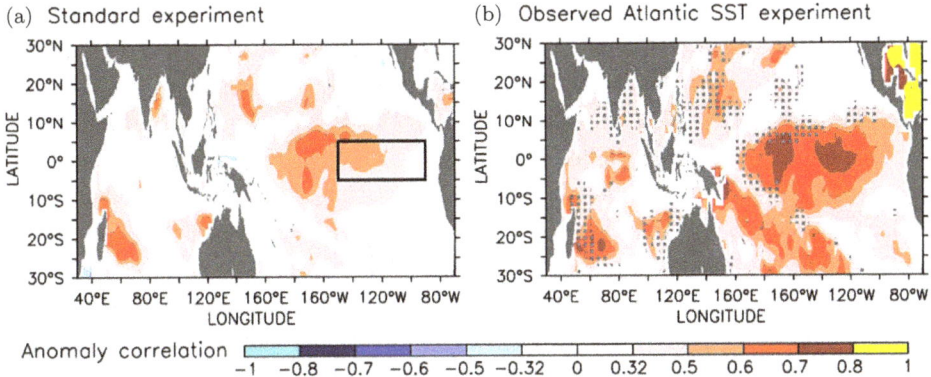

Figure 7.5.   Anomaly correlation skill for October–December average SST for ensemble predictions starting on 1 February performed with (a) a fully global CGCM, and (b) the same model except for Atlantic SST restored to observations. The prediction period is 1980–2005. In (b), regions in color and non-stippled indicate where using observed Atlantic SST leads to a significant increase in skill at the 5% level. (Adapted from Figure 1 in Keenlyside et al. [2013].)

in boreal spring to summer have shown to precede El Niño events in the Pacific by 2–3 seasons through perturbations in the Walker Circulation [An et al., 2021]. This lagged relationship has suggested that Atlantic Niño variability is an important source of additional ENSO prediction skill. Work with different models has provided general support for this notion [Jansen et al., 2009; Frauen and Dommenget, 2012]. Accordingly, anomaly correlation skill in predicting October to December SST from 1 February increases when the climate model predictions include the observed evolution of equatorial Atlantic SST (Fig. 7.5). Keenlyside et al. [2013] indicate that the average skill increase is largely due to the better predictions of the major 1982/83 and 1997/98 El Niño events, achieved through more realistic westerly wind anomalies over the central Pacific during boreal spring. Further work on this aspect of skill increase is required because the impact of the Atlantic Niño on ENSO has varied in time; for example, the impact was weak during the middle of the 20th century [Martín-Rey et al., 2015]. A more detailed discussion of this topic is in Keenlyside et al. [2021].

7.5   Long-term predictions

CGCMs are indispensable tools for predicting and studying a variety of atmospheric phenomena from day-to-day weather changes to long-term climate variations. If an uncoupled AGCM is to be used for NWP, then surface conditions have to be prescribed from some additional source. An obvious first choice has been to keep the SST and sea-ice state constant during the forecast or taken from an observed climatology. One can ask to what extent in time can interactions with the ocean be neglected. The answer is a resounding "no" in reference to timescales beyond the medium range (about 10 days). It has been recognized that the ability to predict climate on seasonal timescales arises from the interaction of the atmosphere with the slower varying components of the climate system [Keenlyside et al., 2021]. The case for seasonal forecasts has been made by the usefulness of seasonal forecasting in the tropics. Predictions of ENSO using both dynamical and statistical models have been examined in Ch. 3. This usefulness has allowed for development of a system for seasonal forecasts using an AGCM with boundary conditions over the tropical oceans provided by other predictions systems. Such a methodology is referred to as a "two-tier forecast system". The reader is referred to Misra et al. [2013] and Li and Misra [2013] for examples of this methodology and its performance.

7.5.1 *Seasonal predictions*

Figure 7.6 shows the mean square skill score (MSSS) of CGCM-based systems at ECMWF for prediction of NINO3.4 SST. This is a critical test of model performance at seasonal time scales in which ENSO is the biggest source of predictability. MSSS is essentially the Mean Square Error (MSE) of the forecasts compared to the MSE of climatology for the location selected. (Here "error" refers to difference with the observation; a "perfect" forecast has a MSSS = 1.) Note the dotted-dashed line labeled "persistence" that corresponds to predict the same initial conditions for all future times. One way to look at the results displayed in this type of figures is fixing a value for the MSSS and comparing the length of lead time required to reach the selected value by different prediction

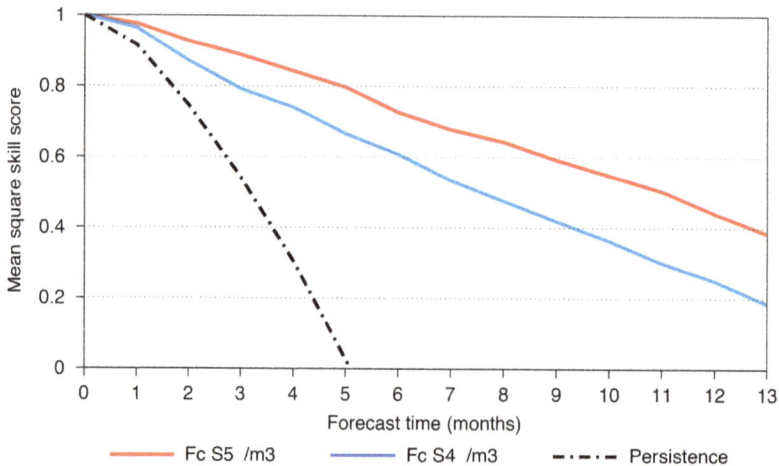

Figure 7.6.   Mean square skill score against climatology for S5 (red) and S4 (blue) forecasts of NINO3.4 SST (4 forecast starts per year, 1981–2016). (Figure 2 in Stockdale et al. [2018]. ECMWF Technical Memorandum No. 835.)

systems. In Fig. 7.6, for example, the lead times at which MSSS for NINO3.4 SST drops below 0.8 is 5 months for system SEAS5, 3 months for system S4 and 2 months for persistence. Stockdale et al. [2018] documents that achieving this skill increase required a multi-year effort from one of the leading operational prediction centers in the world.

7.5.2 *Multi-year predictions*

For predictions in this range of timescales the radiative effects associated with the changing composition of the atmosphere (i.e., by increased greenhouse gas emissions) on the slower varying components of the climate system become important [Meehl et al., 2009]. An important source of predictability lies in the decadal modes of climate variability as the contribution to skill by the initial conditions decreases in time [Hazeleger et al., 2015]. In the Pacific, the main mode of SST variability in the Pacific Ocean on decadal or interdecadal timescales is the Pacific Decadal Oscillation [PDO; Zhang et al., 1997] or the Interdecadal Pacific Oscillation [IPO; England et al., 2014]. In the Atlantic, the main mode of SST variability on multidecadal timescales is the Atlantic Multidecadal Variability (AMV), also known as Atlantic Multidecadal Oscillation [AMO; Kerr, 2000; Knight, 2005]. The AMO also seems to play an important role in modulating the magnitude of global warming through changes in the amplitude of the trends [Ting et al., 2009].

The possibility of predicting phase transitions of the decadal modes of variability has been suggested. Skill in predicting AMV translates to skill in predicting societally relevant quantities, including rainfall over the Sahel [Sheen et al., 2017; Yeager et al., 2018; Mohino et al., 2016] and the intensity of the Atlantic hurricane season [Caron et al., 2017]. Skill in predicting Sahel rainfall is degraded by inaccurate prediction of both PDO and impacts of global warming [Mohino et al., 2016].

Figure 7.7 shows large values of skill over the entire Atlantic and Indian Oceans, and western and South Pacific. There are also indications that initialization of upper ocean heat content can substantially contribute to enhance prediction skill in the North Atlantic and around Antarctica. For a further discussion on the importance of initialization in long-range prediction see Corti et al. [2015].

7.5.3 *Predictions of ENSO changes with global warming*

The way in which ENSO, the strongest year-to-year fluctuation of the climate system, will respond to greenhouse warming remains an open question. For example, IPCC AR5 based on the Coupled Model Intercomparison Phase 5 (CMIP5) concluded that future changes in ENSO intensity are model dependent

(a) **Skill in predicting SST for years 5 to 9**

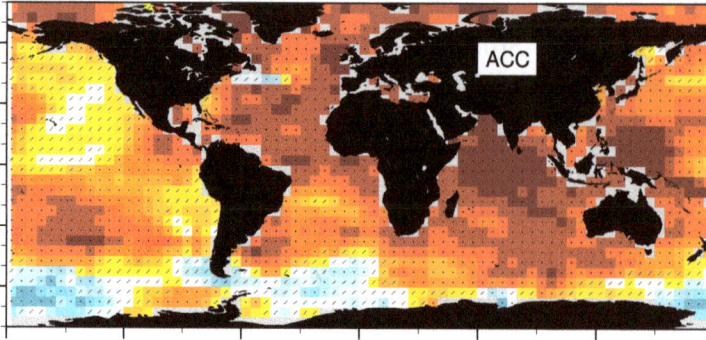

(b) **Skill due to ocean initialisation**

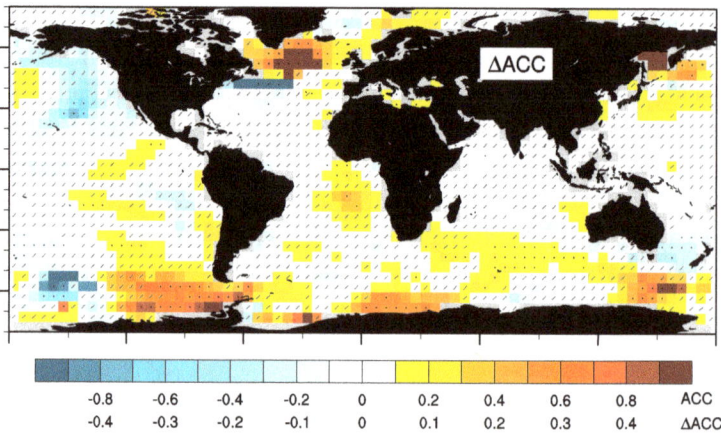

Figure 7.7. (a) Anomaly correlation skill in predicting SST for years 5 to 9 from the CESM Decadal Prediction Large Ensemble experiment that consist of 40 distinct member ensemble predictions started on 1 November every year from 1954 to 2008. (b) Anomaly correlation skill resulting from initialization of the ocean. Note the two different color scales for (a) and (b). Boxes without a gray slash are significant at the 10% level; dots further indicate points whose p values pass a global (70°S–70°N) field significance test. (From Figure 2 in Yeager et al., 2018, © American Meteorological Society. Used with permission.)

and not significantly different from natural modulations, even though the realism of the representations of ENSO in climate models is increasing [Collins et al., 2010; Watanabe et al., 2012; Christensen, et al., 2013]. Part of this uncertainty is associated with the lack of inter-model agreement on the change of tropical Pacific mean conditions in a warmer climate [e.g., Xie, 2020]. An El Niño-like change in a warmer climate has been attributed to a stronger cloud-albedo feedback operating in the eastern Pacific where surface waters are colder in reference to the west; a shallow oceanic mixed layer in the east resulting in an eastward SST response to the surface heating; a weaker evaporative damping over the lower SSTs in the east [Xie et al., 2010]; and a slowdown of the Pacific Walker circulation resulting in a weaker response in global precipitation compared to atmospheric moisture change [Held and Soden, 2006; Vecchi and Soden, 2007] and/or Indo-Pacific warming effects [Tokinaga et al., 2012]. In turn, a La Niña-like change can be driven by the ocean dynamical thermostat [Clement et al., 1996], as the damping effect of mean equatorial upwelling on the SST anomalies strongly compensates the global warming, especially over the eastern equatorial Pacific [Seager et al., 2019]. However, the upwelling cooling effect can be also regulated by the mean climate conditions and its efficiency may decrease as global warming intensifies. Thus a La Niña-like change gradually can become an El Niño-like change under the stronger warming conditions [An et al., 2012; An and Im, 2014].

CGCMs have been applied to estimate changes in extreme ENSO events in warmer conditions (Wang B. et al., 2019). Extreme ENSO events were identified based on rainfall anomalies and meridional SST gradients in the Niño3 region (5°S–5°N, 150°W–90°W) [e.g., Cai et al., 2014]. Simulations by both CMIP3 and CMIP5 obtained that the frequency of extreme El Niño events doubles in response to greenhouse warming [Cai et al., 2014], which arises from El Niño-like SST changes in mean conditions [Xie et al., 2010; Tokinaga et al., 2012]. CMIP5 also obtained that the frequency of future extreme La Niña events also increases from one in every 23 years to one in every 13 years. Such a feature appears due to an increased zonal surface temperature contrast between the Maritime continent and central Pacific, as well as to an enhanced upper ocean vertical temperature gradient [Cai et al., 2015]. Most extreme El Niño events are followed by extreme La Niña events.

El Niño events have been classified into two "flavors" based on whether the maximum SST anomalies are in the eastern or central Pacific (EP and CP El Niño, respectively) [Ashok et al., 2007; Yeh et al., 2009]. CMIP3, 5 and 6 obtained that CP El Niño will be stronger and occur more frequently under global warming [Yeh et al., 2009; Jiang et al., 2020] because the equatorial mean thermocline becomes shallow over central Pacific in association with a weakened Pacific Walker circulation [Vecchi and Soden, 2007]. A consensus on the increased CP El Niño frequency remains elusive, however [Taschetto et al., 2014; Chen et al., 2017; Xu et al., 2017]. An increased variability of EP El Niño under greenhouse warming condition has also been claimed due to enhanced ocean–atmosphere coupling [Cai et al., 2018].

7.5.4 *Climate change predictions*

For longer time scales, multi-model CGCM ensembles of simulations performed under projects such as CMIP3, CMIP5 and CMIP6 [Eyring et al., 2016] provide projections on the climate of the late 21st century. These model integrations are organized by the United Nations Intergovernmental Panel on Climate Change (IPCC). The CMIP projects have tried to involve as many models as currently used around the world. The projections also include uncertainty estimates. Various possible emission pathways have been developed in order to provide a guidance for the future composition of the atmosphere. As part of CMIP6 a number of "Shared Socio-economic Pathways" have been developed covering plausible but different socio-economic, political and technological futures [O'Neill et al., 2017].

The climate projections made have demonstrated that CGCMs succeed in representing the evolution of the global climate since the end of the Industrial Revolution (see Fig. 7.8). There are enough reasons to expect a comparable level of success in predicting the future evolution of global climate although there are additional formidable challenges. One of these challenges is that emissions and other such aspects that are influenced by human intervention also have to be predicted. Another challenge is the systematic errors described earlier in this chapter.

There is another kind of information that CGCMs are in a unique position to provide. Despite a globally uniform increase in the concentrations of emitted greenhouse gases, radiatively forced surface warming can have significant spatial variations. These define climate change patterns that depend on preexisting climate states and through atmospheric and oceanic dynamics can drive changes of the hydrological cycle with global feedbacks. Figure 7.9 shows a prediction of

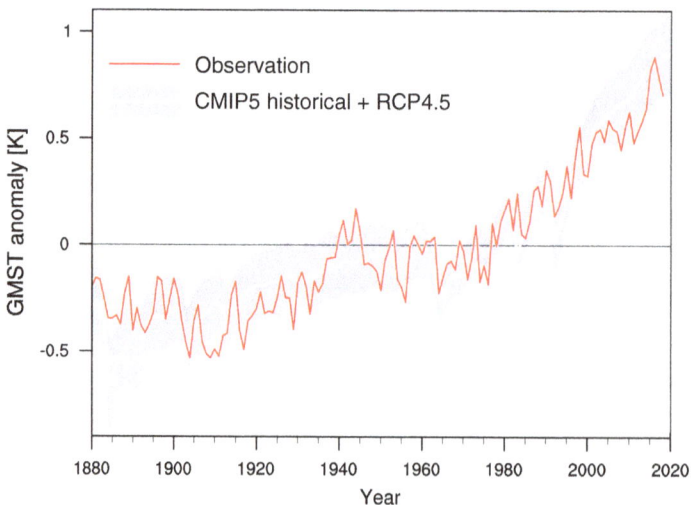

Figure 7.8.   Differences in global mean surface temperature anomalies in references from 1961–1990 from observations and CMIP5 simulations by 37 models using the RCP4.5 scenario. (Figure 6 in Ma et al. [2020].)

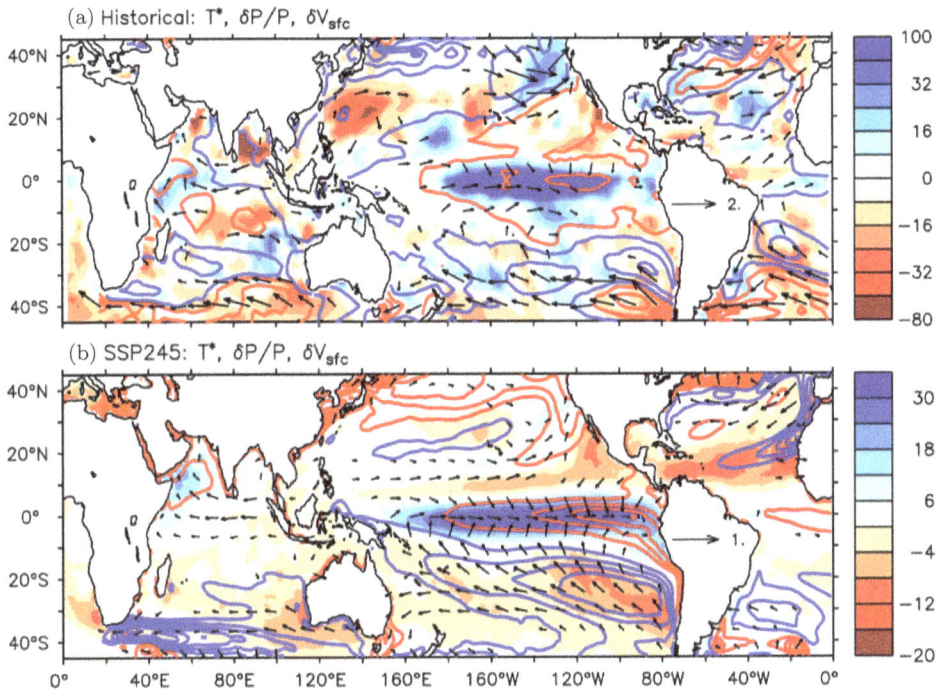

Figure 7.9. Annual mean changes of SST (contours, positive red and negative blue, K), 1000 hPa winds (vectors, m/s), and precipitation (color shading, %) by 37 CMIP6 models. The forcing scenarios include (a) Historical; 1901–1910 minus 1991–2000, and (b) SSP2-4.5; 2091–2100 minus 2015–2024. All changes are normalized by the tropical (20°S–20°N) mean SST warming. (Figure 7 in Ma et al. [2020].)

the geographic variability of expected warming patterns. Much further research is needed on the importance of spatial variations in surface climate change.

7.6 Perspectives

Our emphasis in the present book has been on the simulation and prediction of the large-scale components of climate. In this context, our examination of CGCMs is justified. However, the functions of society demand reliable prediction of an extraordinarily broad range of important weather and climate events. This demand has motivated the development of models for particular regions on Earth (i.e., regional models). Some of the models mentioned in previous chapters are regional because they are focused on ENSO prediction and the ocean component is restricted to the tropical Pacific. Such a methodology required imposing oceanic boundary conditions that basically consisted of relaxation towards observed climatologies. Other regional atmospheric models have been developed that are as complex as AGCMs, both in numerical schemes and physical processes, but with increased resolution at the expense of limiting the application domain in view of limitations in computer resources. Running regional models requires specification of boundary conditions, which may eventually have to be provided by other models with a larger domain such as CGCMs.

A more drastic strategy for operational centers is to adopt a "seamless" approach, according to which configurations of models used for prediction are kept as consistent as possible across a range of applications and timescales [Hurrell et al., 2009; Hazeleger, 2010; Brown et al., 2012]. This brings additional challenges due to the different demands by different applications as the demand to properly simulate different physical processes may not be the same at different timescales. Nevertheless, there is no question that society will benefit from high resolution information on ocean variability to be provided by coupled atmosphere-ocean models, and particularly by CGCMs. For example, having detailed information on SST may contribute to the more successful prediction of a wide range of important events ranging from tropical cyclones and hurricanes tracks [Lloyd and Vecchi, 2011; Morgensen et al., 2017] to large scales such as droughts and floods [Schubert et al., 2016].

Chapter 8

Coupling Software and Technologies

8.1 Historical overview

In addition to ocean and atmosphere, current climate models generally incorporate other components such as land, ocean surface waves and sea-ice models. In such models with multiple components, data calculated in one component is used as boundary condition for one other. For example, the ocean model typically calculates the Sea Surface Temperature (SST) that is needed by the atmosphere model as lower boundary condition for its temporal integration and, vice versa, the atmosphere calculates heat fluxes used as boundary condition by the ocean.

Building a coupled system from individual components needs to meet a variety of scientific, technical, and numerical constraints, such as conservation of physical quantities (water, energy, momentum), numerical stability of the coupling exchanges and representation of motions at several space and time scales. Moreover, computational efficiency of the entire coupled system and of the coupling *per se* is required, as the component models themselves are usually parallel codes running on high-performance parallel hardware.

The software that links together the components of a coupled application is referred to as "coupler", "coupling software", "coupling technology" or "coupling infrastructure". Diverse couplers that carry out similar functions are developed and used in the geophysical community [Valcke et al., 2013]. The main functionality of the couplers is to manage exchanges of coupling fields between component models. As these components usually run on different numerical grids, the coupler needs to regrid the coupling data (e.g., the SST) to transform it from the grid of the source model (e.g., the ocean) to the grid of the target model (e.g., the atmosphere). Furthermore, the coupling software typically coordinates the execution of model components and may provide other support utilities, such as calendar management, logging, and error handling.

The first pioneer atmosphere-ocean coupled simulations at the US National Atmospheric and Oceanic Administration (NOAA) Geophysical Fluid Dynamics

Laboratory (GFDL) did not use specific coupling software [Manabe and Bryan, 1969; Manabe et al., 1975]. The atmosphere and ocean models were run separately. Surface fluxes calculated during the atmosphere run were saved and applied as boundary conditions in the following ocean run. Conversely, the SST computed by the ocean during its run was saved and used as lower boundary condition for the following run of the atmosphere.

A similar methodology was used at the National Center for Atmospheric Research (NCAR), while taking the seasonal cycle into account [Washington et al., 1980]. In this case, the atmospheric model was integrated to provide samples of January, April, July and October fluxes, which were then used to run five years of the ocean and sea-ice models. The new ocean surface temperature and sea-ice distribution were then used as lower boundary conditions for the atmosphere to produce new seasonal flux samples. This asynchronous coupling was iterated a number of times until an approximate equilibrium was reached. The performance of asynchronous coupling methods in two coupled atmosphere-ocean climate models with and without a seasonal cycle was tested by Schneider and Harvey [1985] and Harvey [1986], respectively.

No specific coupling software, except for high-speed network communications, was either used in the first "distributed" coupling performed at the University of California, Los Angeles (UCLA) [Mechoso et al., 1993; 2000]. In this instance, the Parallel Ocean Program (POP) was run on a CRAY at the San Diego Supercomputer Center (SCSC) in La Jolla, CA, USA, and the UCLA AGCM was run sequentially on another Cray at NCAR in Boulder, CO, USA, about 1,750 km away, while exchanging coupling data over the NFSnet high-speed network. Additional experiments were performed with other computer technologies and other locations, both with public and dedicated networks. Further work under the US National Aeronautics and Space Administration (NASA) High Performance Computing and Communication Program (HPCC) led to development of a Distributed Data Broker [Drummond et al., 2001, Sklower et al., 2002].

This DDB supported the data exchanges among the UCLA Earth System Model (ESM) parallel components, which were the UCLA AGCM, POP, UCLA atmospheric chemistry and transport model, and a simplified version of the NASA JPL (Jet Propulsion Laboratory) ocean biogeochemistry model. The components could be distributed on different computing platforms together with data archives and visualization clients. The DDB was implemented as a library used by the components, which remained separate executables, and as one additional specific process. Initially, this process was acting as a registration broker that matched offers to produce data expressed by some components with requests to consume data posted by others, providing technical details for the parallel exchanges. A regridding module was included in DDB in order to support components with different numerical grids. After the initialization, the coupling exchanges happened as specified, directly between parallel models using Parallel Virtual Machine (PVM, http://www.csm.ornl.gov/pvm/).

The development of a generalist coupler in Europe started in 1991 at CERFACS (Centre Européen de Recherche et de Formation Avancée en Calcul Scientifique) with the OASIS (Ocean Atmosphere Sea-Ice Soil) coupler. The first version of the OASIS coupler, OASIS1, released in 1993, was used in a 10-year atmosphere-ocean simulation of the tropical Pacific [Terray et al., 1995]. This first version ran as a separate process through which the coupling fields transited for regridding (Fig. 8.1a). Coupling fields were exchanged through ASCII files and the synchronization was ensured by signals sent to Unix named pipes, a technique for inter-process communication.

Additional modularity was introduced with OASIS2, which could handle an arbitrary number of component models, coupling fields and transformations. OASIS2 API (Application Programming Interface) was parallel in the sense that each process of a parallel model could send or receive only its part of the coupling field (Fig. 8.1(b)). Between 1996 and 2000, alternative communication

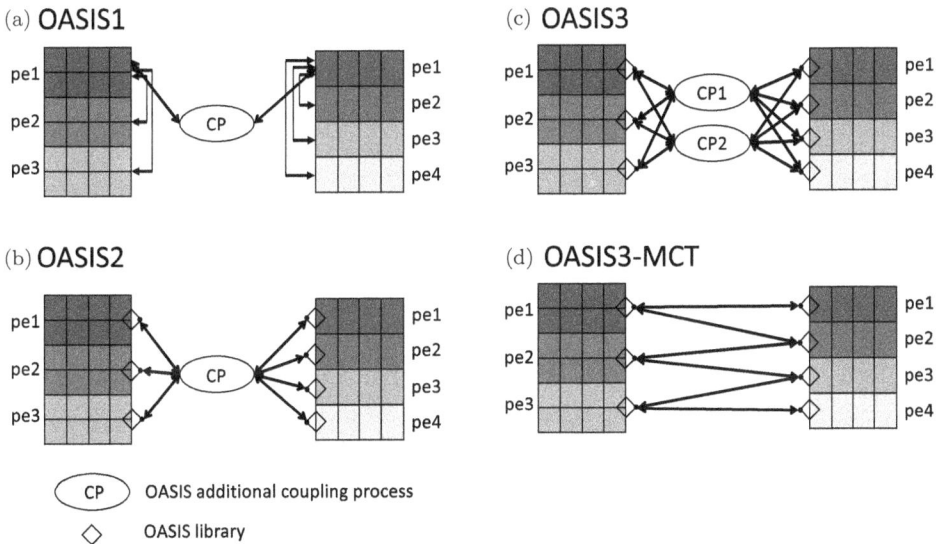

Figure 8.1.   OASIS1, OASIS2, OASIS3 and OASIS3-MCT architecture. (a) In OASIS1, only the master process of each component communicates with the central coupling process, which performs the regridding of the coupling fields; (b) in OASIS2, all processes of each component communicate in parallel with the central coupling process; (c) in OASIS3, there are multiple central coupling processes but each coupling field is entirely handled by only one central process; (d) in OASIS3-MCT, the parallel coupling library linked to the component models performs the communication and the regridding of the coupling fields in a fully-parallel way.

protocols ending with the currently used Message Passing Interface (MPI, https://www.mpi-forum.org) were introduced in OASIS2.

In 2003, the development of a new Application Programming Interface (API) lead to the OASIS3 release [Valcke, 2013]. OASIS3 was both a library linked to the component models and a central coupling process. Multiple central coupling processes could also be used, each process handling one or more than one entire coupling fields (Fig. 8.1(c)); the parallelization of this implementation was therefore, by construction, limited by the number of coupling fields. The first widely used parallel coupler in the OASIS series was OASIS3-MCT [Craig et al., 2017], of which version OASIS3-MCT_4.0 was released in 2018. OASIS3-MCT is described in Sec. 8.2.1.

The CPL3 coupler developed at NCAR was included in the first public release of the Community Climate System Model (CCSM) in 1996 but was used internally at NCAR as early as in 1994. CPL3 was a sequential hub-and-spoke unit organizing the coupling exchanges and executing coupling operations such as regridding, merging, or flux calculation. The hub-and-spoke design implies that all coupling requests and offers are addressed by the components to a central hub that makes the links between the requests and the offers and coordinates the data exchanges between the components. CPL3 was a process distinct from the component models running concurrently on disjoint sets of hardware processors. The first parallelism of the coupler based on OpenMP was introduced in CPL5 used in CCSM2 around 2004. The CPL6 coupler [Craig et al., 2005] was later fully parallelized using MPI, both for the coupler operations and for the communications. The CPL7 coupling architecture [Craig et al., 2012] developed for CCSM4 and for the Community Earth System Model 2 (CESM2) took a completely new approach with respect to the high-level design of the coupled system. CPL7 follows what we call the "integrated coupling approach" (see Sec. 8.3) in which the coupling software ensures, within one executable, the execution of the components and the exchange and remapping of coupling data between these components. CPL7 is described in more detail in Sec. 8.2.5.

The Parallel Climate Model [PCM; Washington et al., 2000] supported by the US Department of Energy (DOE) also followed the integrated coupling approach. Within a single executable, a driver calls the components via subroutine interfaces with sequencing and coupling operations tailored to particular scientific needs. A flux module connected the PCM components ensure calculation of heat, water, and momentum fluxes while conserving the total global energy of the system.

During the same period, GFDL undertook a modernization of its climate codes that resulted in the GFDL Flexible Modeling System (FMS). A main design principle of FMS was to provide a standard programming layer to abstract away the details of the underlying computing architecture in order to facilitate sharing of code and development costs across multiple institutions.

FMS provides the "plumbing" for handling the distribution of the work in parallel among the available computing resources and for data transfer between model components forming an Earth System Model. FMS defines also standard interfaces to build a single-executable ESM based on interchangeable components. More details on FMS are provided in Sec. 8.2.4.

The ESMF collaboration (https://earthsystemmodeling.org) development started in the US in the early 2000s with the Common Modeling Infrastructure Working Group (CMIWG), an unfunded collaborative effort for Earth system model development in the US. In 2000, the NASA Earth Science Technology Office (ESTO) called for the creation of a common framework for Earth System Modeling. In the first period of this effort, a prototype of the framework was developed [Hill et al., 2004; Collins et al., 2005]. The ESMF was used at NASA Goddard to construct a new model, the Goddard Earth Observing System (GEOS). In 2008, the US National Weather Service, Navy and Air Force started the National Unified Operational Prediction Capability [NUOPC; see http://earthsystemmodeling.org/nuopc/; Theurich et al., 2016], a joint multi-agency project aimed to increase the level of interoperability between ESMF components. Currently, ESMF is governed and further developed by a set of partners in the US that includes NASA, NOAA, Department of Defense (DoD) and National Science Foundation (NSF). The concept underlying ESMF is that complex applications can be built up based on coherent modules, or components, adapted to expose standard data structures and called via standard interfaces. ESMF is described in more detail in Sec. 8.2.2.

8.2   Coupling software used in contemporary atmosphere-ocean coupled models

The brief historical overview in Sec. 8.1 highlights that developers of coupled systems and coupling software face many choices of implementation. This section describes in detail the couplers mostly used at the present time by the climate modeling community.

To recapitulate, a model in which components require information from each other can run as one executable integrating all components or as a system with each component being a separate executable. The components themselves can run sequentially, concurrently, on in a mixed mode and it is very often difficult, if not impossible, to define a layout that optimizes at the same time the load balancing, the use of computing resources and the throughput of the coupled model. A top-level driver can explicitly manage the execution of the components, or the components can implicitly synchronize themselves via the coupling exchanges. These exchanges may take place directly between the components or through a central hub that receives and dispatches the coupling fields. The components supported can be parallelized with different standards

(MPI, OpenMP, mixed) and be coded in different languages (Fortran, C, C++, Python).

Regridding algorithms acting on different types of grids can be supported using weights that can be generated offline before the run or calculated online by the coupler and calculations can be done once at initialization or dynamically during the run to support grids evolving with time. The coupling software can offer only coupling services or can be developed as a complete infrastructure managing all technical aspects of the coupled model such as the component internal parallelism, I/O, calendar, etc. The number of functionalities that can be supported is almost boundless and in practice, even if the couplers have become more and more sophisticated over time, no coupler may be able to address all of them. As a consequence, choices are made, primarily dictated by the needs and resources of the user community.

In the following, generic coupling software used in atmosphere-ocean models are presented. By "generic", we mean "intended to couple any component" and not hardwired to one specific coupled application. OASIS3-MCT and ESMF are discussed in greater detail as they are certainly the most widely used at the present time. Other coupling software, which either influenced the development of atmosphere-ocean models such as MCT, FMS and CPL7 or which emerged more recently, such as YAC [Hanke et al., 2016], C-Coupler2 [Liu et al., 2018] and MOAB-TempestRemap [Mahadevan et al., 2020] are also presented.

8.2.1 *OASIS3-MCT*

OASIS3-MCT is the latest stage in the evolution of the OASIS1, OASIS2 and OASIS3 series towards a fully parallel coupler in response to the ever increasing needs of the user community. Low-intrusiveness in the component codes was, and still is, a major design principle. OASIS3-MCT retains the already parallel API of OASIS3 but replaces the underlying communication layer with MCT. OASIS3-MCT runs as a parallel library linked to the component models and manages exchanges and regriddings directly between them without going through any central process (Fig. 8.1(d)). MCT data types are hidden from the user and standard Fortran data (real or integer) are passed through the API. There is no driving layer and so the user has to take care that some global parameters (e.g., the total run duration or the calendar) are defined coherently across component models.

As full backwards compatibility was ensured between OASIS3-MCT and OASIS3 APIs, OASIS3-MCT was rapidly adopted by the user community. The high level of support offered by the developers is one of the key reasons for its popularity today. As of 2021, OASIS3-MCT is used by 67 modeling groups to assemble more than 80 different coupled applications worldwide. These applications include ocean and atmosphere models but also sea-ice, sea level, surface ocean wave, ocean biogeochemistry, land, vegetation, river routing, hydrological and atmospheric chemistry models in global or regional configurations. OASIS3-MCT is also used to exchange information with auxiliary modules, such as file readers. Major climate modeling groups in Europe use OASIS3-MCT, in particular in their coupled model developed for the 6th phase of the Coupled Model Intercomparison Project (CMIP6).

The component models coupled with OASIS3-MCT remain individual executables in the UNIX sense with their main characteristics, such as internal parallelization or I/O, untouched with respect to their standalone mode. To interact with the other components of the coupled system, the component models need to call the routines of the OASIS3-MCT API to perform the different coupling steps described in the next paragraphs.

8.2.1.1 *Initialization and definition*

All component processes first need to initialize the coupling. OASIS3-MCT creates for each component a specific local MPI communicator gathering all its parallel distribution, to be used by the component for its own internal parallelization.

The coupling fields of a component model are usually computed in parallel by its different processes. Using OASIS3-MCT, each process can send and receive only its part of the coupling field. To do so, each process describes its local partition with a vector of integers describing the offset and the extent of the local partition in a global index space covering the whole component grid. Any type of parallel distribution can be expressed, either as an ensemble of discontinuous segments, or with an explicit list of global indices associated with each local grid point.

OASIS3-MCT reads the information needed for regridding of the coupling fields (e.g., longitude and latitude of the grid points and grid cell vertices, the grid point mask, and the surface of the grid cells) from NetCDF (Network Common Data Format) auxiliary grid data files. These files can be created either by the user before the run or produced by the components by calling the appropriate API routines with arguments containing the grid information.

Each component process then declares the coupling fields that will be sent or received during the simulation, providing the different characteristics of the field such as its coupling symbolic name and its status (sent or received). Coupling field exchanges are specified in the *namcouple* configuration file (see Sec. 8.2.1.2) by associating the source and the target symbolic names of a coupling field. OASIS3-MCT also calculates the weights and addresses needed for the regridding of the coupling fields or reads them from files created offline by the user. Based on this information, OASIS3-MCT defines the communication patterns between the source and the target parallel distribution that are used at runtime for the coupling field exchanges.

8.2.1.2 *Coupling send and receive*

The OASIS3-MCT sending and receiving routines are based on MPI. The field arrays can be 1D or 2D and can also be bundled, i.e., have an extra non-spatial dimension, for example for a field covering different ice categories.

A source component does not know to which other component the coupling field is sent to and a target component does not know where the coupling field comes from. The match between a sending action (put) and a receiving action (get) is activated by OASIS3-MCT given the specifications provided by the user in the *namcouple* file for each specific run. The user may also specify in the *namcouple* that the source of a get or the target of a put is a NetCDF file. In this case, OASIS3-MCT reads the field from a file or writes it to a file. This allows an easy switch from the coupled to the forced mode, as input fields can then come from forcing files.

The put and get can be called at each time step anywhere in the component code. The time at which the call is valid must be given as argument. OASIS3-MCT analyses this time and the put or the get is actually performed only if the time corresponds to a coupling period, specified in the *namcouple*. Therefore, changing the coupling period is simply achieved by changing the coupling period in the *namcouple* without any modifications in the component codes. The get is blocking, i.e., it will return only when the coupling data is effectively received, but the put will return even if the exchange is not completed after storing the coupling data into MPI buffers.

It is also possible to match a put performed at a certain time with a get performed at a later time by specifying a LAG in the *namcouple*. In that case, the put is actually performed when the time + LAG equals a coupling period and it then matches the get performed at the resulting later time.

Exchanging multiple fields as a single coupling operation, which can improve performance, is also possible for fields for which coupling options are identical. The get and put calls in the model are still individual, but OASIS3-MCT aggregates the corresponding coupling exchanges into a single step.

8.2.1.3 *Communication and component layout*

Internally, parallel distributed exchanges and parallel regridding are managed by MCT. For regridding, the weight files can either be pre-defined and read in directly or generated in the initialization phase. The regridding *per se* can be done either on the target processes or source processes, as specified by the user in the *namcouple*.

OASIS3-MCT supports traditional coupling between separate executables running concurrently as separate executables on different processes. It also allows for coupling exchanges within a single executable between components running concurrently or sequentially, on overlapping or partially overlapping processes. This provides significant flexibility to layout the component models on parallel tasks in relatively arbitrary ways to optimize the performance.

8.2.1.4 *Configuration*

The configuration of the coupling exchanges performed by OASIS3-MCT for each specific run of a coupled model must be provided by the user in a text file, the so-called *namcouple*. The first part of *namcouple* is devoted to general coupling parameters such as the number of coupling fields, the total run time or the debug level. The second part provides specifications for each coupling exchange. In this part, the symbolic name specified in the source component and the symbolic name specified in the target component are associated, telling OASIS3-MCT to make the link between put and get calls in the source and target component models, respectively. For each coupling field, the user also specifies the coupling period, the lag if any, and the list of transformations including the regridding.

8.2.1.5 *Regridding and other transformations*

The first phase of the regridding is the generation of addresses and weights that describes which points in the source grid contribute the calculation of each destination grid point value and their respective weighting factor. The second stage is the multiplication of values on the source grid by these weights to produce values on the destination grid.

The generation of the regridding addresses and weights within OASIS3-MCT is performed by a parallel version of the Spherical Coordinate Remapping and Interpolation Package (SCRIP) library [Jones, 1999] for 2-D fields on the sphere. Algorithms include n-nearest-neighbor (possibly Gaussian-weighted), bilinear, bicubic and first and second order conservative remapping. If the user provides a file with pre-calculated weights, OASIS3-MCT reads and can use them to regrid any type of 1-D, 2-D, or 3-D fields, using a 1-D degeneration of the grid structure. OASIS3-MCT does not support dynamic grids.

OASIS3-MCT also offers time accumulation, averaging, minimum or maximum. These operations are performed on coupling field arrays provided as argument of the different put calls over the coupling period and the resulting field is finally sent at the coupling frequency. Addition or multiplication by a scalar are also supported. Finally, OASIS3-MCT also offers forced global conservation of the coupling fields. See Sec. 8.3.1 for explanations on those transformations.

8.2.1.6 *Performance*

Different tests were realized on different supercomputers with a "toy" coupled model to evaluate the performance of OASIS3-MCT at high number of cores.

Toy components are parallel programs that do not include any physics or dynamics, like real geophysical models, but that implement realistic exchanges of coupling fields defined on realistic grids. In these tests, each toy component used a grid with about a million points partitioned on up to 10,000 cores. These components exchanged one coupling field back-and-forth, which was regridded each time.

In that case, the regridding weight calculation remained below 100 seconds. The coupling initialization time, including the reading of the regridding weight files, was at most of the order of one minute. The cost of a back-and-forth coupling exchange, including the communication and the regridding, remained below 0.01 second. The sum of all these costs is certainly much less than the cost of the computations of a real coupled model at this resolution. These tests therefore lead to the conclusion that OASIS3-MCT most likely provides an efficient coupling solution for still many years, even for very-high-resolution parallel coupled models.

8.2.2 *ESMF*

The ESMF is an open source software for coupling model components to form weather, climate, coastal, and other Earth science related applications. The framework allows for a specific part of a model developed by a particular modeling group to become a module of large scientific applications developed either by the same group or by others. One important objective is to limit redundancy in code development and enable components to be exchanged between different modeling centers. The scientist only codes the scientific part of his model into modular components and adapt them to the standard calling interface and standard data structures of the shared infrastructure software. The ESMF software provides the underlying layers necessary for an efficient parallel execution of the scientific applications on different computer architectures allowing for the coupling of the module to other components, including transfer and transformation of the coupling data.

ESMF offers a "superstructure" of functions and component wrappers with standard interfaces which forms ESMF coupling functionality. The superstructure assembles the components interconnecting their input and output data. ESMF also provides a separate "infrastructure" of utilities that can be used internally by the components to handle technical aspects of the coding, e.g., parallelization, time and calendar management, error handling, and I/O. As illustrated in Fig. 8.2, the user code lies between the superstructure and the infrastructure. Note, however, that not using any of the infrastructure does not prevent a component to be coupled through ESMF superstructure. An ESMF coupled application is a hierarchical composition of the superstructure, user components, and the infrastructure when used.

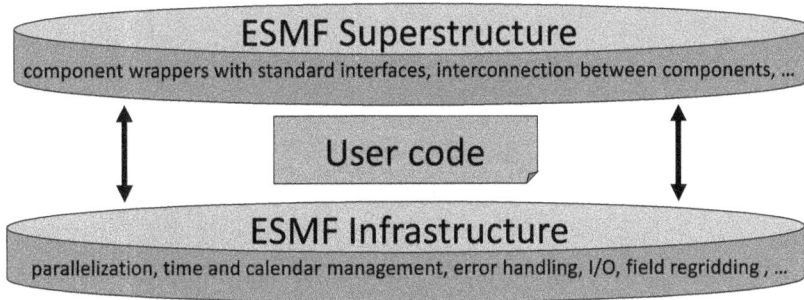

Figure 8.2.   Illustration of the ESMF architecture: the user code lies between an upper-level superstructure layer and a lower-level infrastructure layer.

8.2.2.1 *Using ESMF superstructure to assemble coupled applications*

Two types of components are defined in the ESMF superstructure: "Gridded components" containing the code modeling a physical domain or realizing a computational function, and "Coupler components" transforming and transferring physical fields between Gridded components. Following the concept of modularity, Gridded components internally hold no information about the other Gridded components that they interact with. When needed, they receive this information through their argument list.

ESMF itself does not provide a predefined driving layer for the components but provides a software to build a hierarchical coupled application based on those components, as illustrated in Fig. 8.3. It also offers a full interface to Fortran 90 and partial interface to C/C++ and Python, and supports user codes parallelized with MPI, OpenMP and mixed MPI/OpenMP.

In order to adopt ESMF, users first have to split their original codes into initialize, run, and finalize parts, called "methods". Then, they have to wrap the native model structures of the coupling data with ESMF data structure, transforming the split code into a set of Gridded components. A Gridded component takes in one Import State and produces one Export State, which contain the physical coupling fields to be passed to or from other components. ESMF does not prescribe the units of the fields but proposes mechanisms to describe then as well as their grid coordinates. It is permitted to use pointers to component data for the content of Import and Export State objects in order to avoid copies of potentially large data structures. Ocean, atmospheric, land surface and sea-ice models are typical examples of Gridded components in an Earth System Model. Code for linear algebra can also be wrapped in a Gridded component.

If the NUOPC layer is not employed (see 8.2.2.2), the user has to write or customize existing Coupler components that will regrid, transform or merge the coupling data between the Gridded components. The user also has to write a

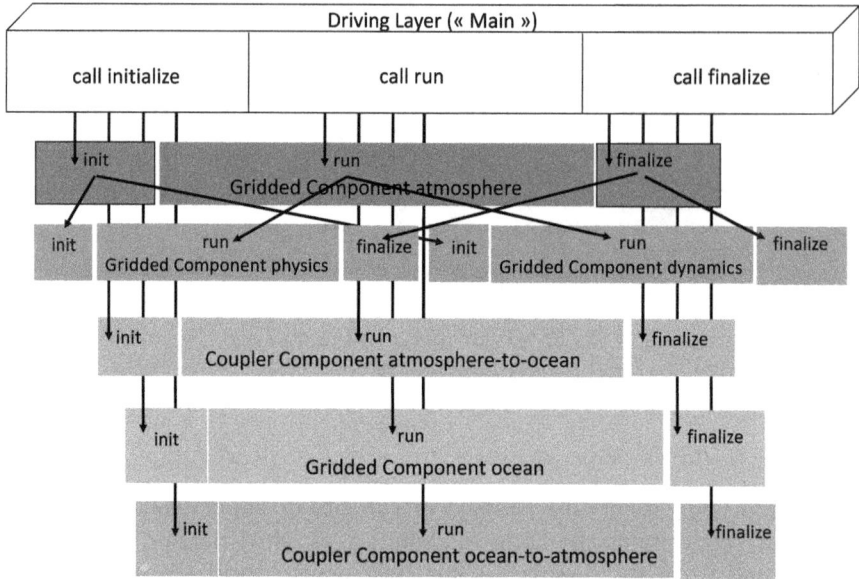

Figure 8.3.   A hierarchical coupled model assembled with ESMF. The driving layer calls sequentially the initialize, run and finalize methods of an atmosphere Gridded component, an atmosphere-to-ocean Coupler component, an ocean Gridded component and an ocean-to-atmosphere Coupler component. In the atmosphere, which is a parent component, these calls cascade to the physics and dynamics child Gridded components.

driving layer that performs all steps required to assemble a coupled application using the Gridded and Coupler components. This driver calls the initialize method of the different components, implements a timestep loop into which it calls the run method of the components, and finally calls the finalize methods of the components.

ESMF can accommodate a wide range of component sequences, running them sequentially (i.e., one after the other on the same set of processors), concurrently (i.e., at the same time of disjoint sets of processors), in mixed mode, or nested into one another.

Usually the resulting coupled application contains all components linked into a single executable program. But ESMF also allows a Gridded component to run as a separate executable or even as a Web service thereby allowing coupling of Gridded component through the Internet.

8.2.2.2 *Interoperability, NUOPC and ESPS*

Adopting ESMF data structure and component wrappers does not guarantee interoperability between ESMF components. In 2007, a consortium of NOAA, US Navy, and US Air Force operational weather prediction centers decided to develop the National Unified Operational Prediction Capability (NUOPC) layer, a set of conventions and generic higher-level templates, to increase interoperability of ESMF components.

The NUOPC layer offers templates for the top driving layer and for Gridded and Coupler components, defining a standard way to access the component coupling fields and clock information. NUOPC also proposes Connectors, i.e., code implementing the communication between components and some field transformations, and Mediators wrapping custom coupling code (e.g., flux calculations or averaging). The ESMF regridding utility (see Sec. 8.2.2.3) can be activated in the Connectors or in the Mediators. A Driver is specialized by harnessing specific NUOPC components, Mediators and Connectors. Examples of architectural options offered by NUOPC combining components, Connectors and Mediators are illustrated in Fig. 8.4.

The NUOPC layer deals with the initialization of the coupled model defining and setting up the coupling exchanges between the components. Typically, the first action for the components is to advertize the fields of their Import and Export States with associated metadata. The generic Connectors use this information to determine which advertized fields should be matched and to construct the connection maps between the components.

For the run itself, the NUOPC layer is capable of implementing different coupling algorithms. Different run sequences are defined, each one being associated with specific time stepping and chain of components, Connectors and Mediators. A Driver can activate multiple run sequences, which in turn

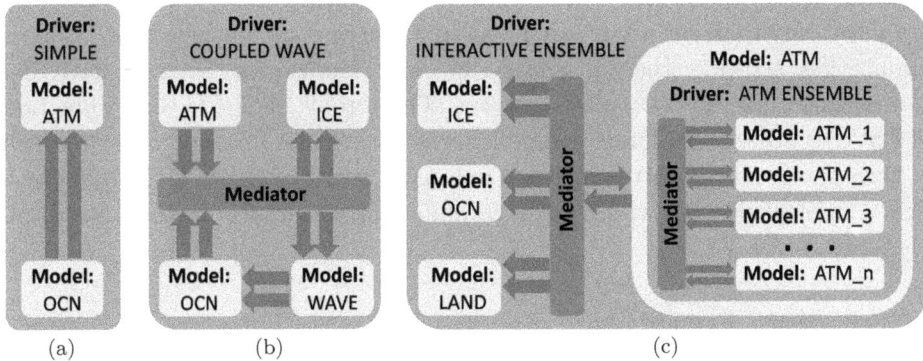

Figure 8.4. Examples of architectural options supported by NUOPC. In (a), an atmosphere is simply connected to an ocean without any Mediator and the regridding takes place directly in the Connector. In (b), the Mediator is used as a central hub between atmosphere, ice, ocean and wave components, but direct connections are established between the tightly coupled ocean and wave components. In (c), NUOPC is used to set up a multi-model interactive ensemble where each member uses a different version of an atmosphere model; two mediators are used, one acting as a hub between the ice, ocean, land and the member-specific atmosphere, and the other staging the multiple versions of the atmospheric model. (Figure courtesy of Gerhard Theurich from NRL/ESMF/SAIC.)

can call other run sequences. A simple run sequence is the one associated with asynchronous coupling between an ocean and an atmosphere (see also Sec. 9.2.1) where each component uses, for a specific coupling period, the coupling fields produced by the other component at the end of the previous period. Another example is a leap-frog run sequence where one of the components receives for each coupling period the coupling fields produced by the other component one coupling period ahead. Chaining different coupling sequences allows the implementation of very complex sequences with components running sequentially, concurrently or in a mix of both modes.

To facilitate even further the development of coupled applications based on ESMF-compliant components, the national Earth System Prediction Capability (ESPC) project coordinates the development and collection of NUOPC-compliant components, the Earth System Prediction Suite (ESPS). ESPS components are tested, documented, versioned and have clear terms of use (e.g., public domain, open source, proprietary status). Different tools exist to facilitate the writing and compliance verification of ESPS components and coupled models. These include the Compliance Checker that analyses each component and checks for the presence of the required initialization, run, and finalization phases, for a correct timekeeping, and for component and coupling field metadata. The Component Explorer is another tool that can be used as a component driving layer to perform different checks, such as whether or not required inputs are provided, thereby offering a way to evaluate the behavior of a single component outside its full coupled application. Finally, the Cupid Integrated Development Environment provides a comprehensive graphical environment that can be used to create an outline of a coupled application showing the initialize, run, and finalize methods of NUOPC components and their compliance status. Those tools facilitate the production of ESPS weather and climate components structured for interoperability, which in turn, favors collaborative development and transfer of research to operations. ESPS implements the vision of a common infrastructure for Earth system modeling in the US.

8.2.2.3 *ESMF Infrastructure*

The ESMF infrastructure defines classes to address technical issues arising when coding a parallel component, and ensures performance over different computer systems. For example, FieldBundle, Field or Array classes contain field data modeled by the component and attributes that provide the data units or memory layout; the Grid class can hold information about the discretization grid onto which the field data is expressed; the Time class can be used to describe specific times during the component run. The infrastructure offers utilities acting on those classes that can be used by the component for time management, message logging, error handling, code documentation, configuration, profiling, I/O, internal parallelization, data regridding, redistribution and communication. Infrastructure utilities can be used independently from the superstructure.

The Time Manager utility can be used for time and date representation, model time advancement, alarm setting and identification of unique and periodic events. It allows Gridded and Coupler components of a coupled application to be latched to a common controlling clock, thereby ensuring their synchronization. Different calendars are supported, including Gregorian, no-leap Gregorian, 360-day, Julian, Julian Day, and no calendar (i.e. the elapsed model time is expressed only in terms of hours, minutes, seconds). Varying and negative time steps are supported.

The infrastructure also defines a set of classes for high performance parallel I/O for storing and retrieving data objects — such as Arrays and Fields — to and from disk storage. The current I/O functionality relies on the Parallel I/O (PIO) library developed by NCAR and DOE laboratories and supports I/O of binary and NetCDF files.

The ESMF regridding utility is a particularly powerful tool. It can be used within an ESMF application, or offline, to perform the calculation of the regridding weight matrix and the regridding *per se*. It is also possible to generate and store the regridding weight matrix for a later use. Adaptive grids are possibly supported in the sense that an ESMF application can be built to recalculate the weights if the target and/or the source grid change during the run. The source and destination fields can be discretized on global or regional domains using logically rectangular grids, unstructured meshes or observational data streams. On these different discretization types, ESMF supports regridding on all combinations of 2D or 3D spherical or Cartesian coordinates with different methods: nearest-neighbor, bilinear, higher order based on patch recovery, first- and second-order conservative (see Sec. 8.3.1). ESMF regridding utility offers a wide range of options for pole treatment, masking and handling of unmapped points. ESMF also includes a Python interface to its regridding utility.

8.2.2.4 *Users and user support*

To ensure the software robustness, a suite of over thousands of tests and examples is scheduled to run automatically every night on many different platforms. These tests show that the grid remapping and parallel communications are fast and scalable.

ESMF/NUOPC is used in major coupled systems at NASA, US Navy, NCAR, and NOAA as well as in other modeling applications from universities and major U.S. research centers. Currently, ESPS gathers more than 30 ESMF/NUOPC-compliant models including atmosphere, ocean, sea-ice, land ice, hydrology, land surface, chemistry/aerosol, ionosphere, and wave components. ESMF grid remapping tool is also being used for data analysis and visualization in packages such as the NCAR Command Language (NCL).

Since ESMF is a complex software offering many functions, users can expect a significant learning curve before mastering the software. Free and active user support is offered to users in order to facilitate this phase.

8.2.3 *MCT*

The Model Coupling Toolkit [MCT, Larson et al., 2005] is a coupling library that can in principle assemble any multi-physics coupled system. MCT imposes no constraint on the application design (number of executables, component decomposition, etc.) as long as the parallel decomposition of the components can be described in a 1D global index space.

MCT design philosophy is based on minimal invasiveness and flexibility, considered vital to the development of long-life-cycle coupled models. MCT proposes a light and low-intrusive Fortran API and bindings for C++ and Python. To use MCT, the developer implements API calls in the component model code. In the initialization phase, the coupling data is declared and its parallel decomposition is described. The most delicate task is to describe the local partition of each process in a global index space representing a linearization of the whole component grid. Based on this information, MCT defines a domain decomposition descriptor (DDD) object based on the 1D global index space. Parallel communication patterns between the components are computed from source and destination DDDs using regridding addresses and weights that have to be precomputed offline with another software. The developer also inserts send and receive calls between the component model pairs to activate parallel coupling field exchanges, explicitly specifying the source of a receive action or the target of a send action. Parallel data transformation and regridding also have to be invoked explicitly through API calls.

MCT and OASIS3-MCT designs are obviously very close. As detailed in Sec. 8.2.1, OASIS3-MCT provides a layer above MCT that facilitates and automates its use based on the coupled model configuration defined for each specific simulation. For example, with OASIS3-MCT, the target of a put or the source of a get respectively are not hard-coded in the component codes; these are automatically defined based on the information provided by the user in the *namcouple* configuration file.

MCT is included in all NCAR couplers since 2004 including CPL7 (see Sec. 8.2.5) used in CESM2 (see Sec. 9.2.4). MCT is also used in the first version of E3SM (Energy Exascale Earth System Model, https://e3sm.org, Golaz et al., 2019). Currently, MCT supports applications on hundreds of thousands of processors and is well positioned for future coupled model challenges. However, exascale platforms will most likely require rewriting key parts of MCT. In particular, the low memory per node expected at exascale will require revisiting the storing of the field data and DDD replications. Work is currently ongoing along those lines with the development of the MOAB-TempestRemap coupler (see Sec. 8.2.8).

8.2.4 *FMS*

The Flexible Modeling System (FMS) is used in the GFDL models developed for CMIP6, GFDL-CM4 and GFDL-ESM4. FMS addresses the need to develop high-performance kernels for low-level numerical algorithms, while offering a high-level structure to build climate models using component models developed by independent research groups. The system handles the data transfer between the component models and their internal parallelization and offers other utilities for I/O, time management, diagnostic and error handling, or for interfacing with scientific software methods such as FFTs.

FMS high-level infrastructure proposes standard interfaces to build single-executable coupled models based on atmosphere, ocean, land and ice components, possibly including also components of the carbon cycle such as atmospheric chemistry, terrestrial biosphere or ocean biogeochemistry. To be included in the system, components must be "wrapped" in FMS-specific data structures and procedure calls. FMS offers some flexibility to swap components; for example, behind an FMS ocean interface may lie a full ocean model, a simpler mixed layer, or even a routine reading appropriate oceanic datasets. The FMS coupled modeling system also includes a parallel ensemble adjustment Kalman filter for data assimilation in which the different members are treated as concurrent components.

To help the problem of consistent flux calculation given different discretization grids in the different CGCM components, FMS proposes the concept of the "exchange grid" [Balaji et al., 2006]. The grid of each component is formed by cells defined by edges joining pairs of vertices. An exchange grid between two component grids is constituted by cells defined by the intersection of all cells of the two parent grids. Each exchange grid cell is therefore uniquely associated with exactly one cell of each parent grid, representing a fraction of the parent grid cells. The exchange grid between atmosphere and land-ice parent grids is illustrated in Fig. 8.5. An ice model always sits on top of the ocean model even when there is no ice, but this has no impact on the exchange grid as the ice and ocean share the same grid.

Coupling fields transferred from one component to the other are first expressed on the exchange grid and are then aggregated on the target grid, considering the fractional area of the corresponding exchange grid cells. For example, a turbulent flux can be calculated for each exchange grid cell based on the exact values of the land surface temperature at its corresponding land cell

Figure 8.5.   The exchange grid (EXC) is defined as the intersection of the atmosphere (ATM) and surface land (LND) and ice (ICE) model grids. The ice and the ocean (OCN) models share the same grid.

and the atmospheric temperature at its corresponding first-level atmospheric cell. The flux for each atmospheric cell is then the aggregation of the fluxes of the corresponding exchange grid cells weighted by their fractional area and similarly for the land and ice models. In practice, only few grid cells of one component intersect cells of the other component and therefore the exchange grid size is of the order of the number of cells of the component grids.

Another key feature of FMS is to allow for an implicit resolution of the vertical diffusion of heat, humidity and momentum across all atmospheric and land or ice vertical levels even if these models are run on different grids. An implicit treatment of the vertical diffusion enhances the stability of the calculation and implies the resolution of a tridiagonal system from the top of the atmosphere model to the bottom of the land/ice model, involving at their interface the exchange grid where the quantities are regridded and where the surface turbulent fluxes are finally computed.

FMS offers state-of-the-art performance as it was shown to scale up to $O(10\ 000)$ processors and has been active for more than two decades. In particular it supports GFDL contribution to CMIP6.

8.2.5   *CPL7*

CPL7 is the most recent coupler developed at NCAR providing the coupling infrastructure for CCSM4 and for CESM2 used in CMIP6 [Craig et al., 2012]. CESM2 is a state-of-the-art global Earth System Model comprising atmosphere, land surface, ocean and sea-ice but also land ice, land and ocean biogeochemistry, and atmospheric chemistry components. As in FMS (see Sec. 8.2.4), the component models can be either active real dynamical components or "data" components, which read coupling data from files. CPL7 therefore allows to run a component stand-alone simply by activating the "data" version of the other components, reading appropriate forcing datasets.

With CPL7, all component models are split in initialize, run and finalize parts, which are merged into one single executable. The execution is coordinated via a top-level driver that runs on all processes and calls the different components through standard subroutine interfaces. The driver also calls a "flux coupler" to interpolate, rearrange or merge fields, calculate fluxes, and generate diagnostics. The sequencing of the driver run loop over the flux coupler, land, sea-ice, atmosphere, and ocean components is hard-wired and is based on scientific constraints first and performance optimization second (see also Sec. 9.2.4).

The driver allows the components and the flux coupler to be flexibly placed on all or arbitrary subsets of hardware processors and to be run sequentially, concurrently, or in a mixed mode, as shown in Fig. 8.6. Each component can use a hybrid mode of parallelism using OpenMP threads within each of its MPI task. The layout is defined for each component by the user through a namelist specifying the number of MPI tasks, the number of OpenMP thread per MPI task and the global MPI rank of the root process. The optimal layout depends on the characteristics of the components (resolution, execution time, scalability, internal load balance, etc.), which may in turn vary depending on the computing architecture.

The objective is to optimize the performance of each component while minimizing the load imbalance. Typically, a series of short test runs is done with different configurations to determine the smallest runtime and a reasonable load balance setup for the production job. In the fictive case illustrated in Fig. 8.6, the hybrid layout (c) is the most efficient one as it produces the fastest runtime and small load imbalance. It is important to stress that the layout of components on processors does not change the model sequencing, coupling algorithm or results.

The communication is ensured by MCT (see Sec. 8.2.3) but alternative component interfaces consistent with the ESMF gridded component specification (see Sec. 8.2.2) are also available. The component data has to be wrapped or copied into specific data types. MCT manages all data exchange, rearrangement and data mapping between component processors, decompositions and grids. With CPL7, the remapping weights need to be precomputed off-line; they are read in and then applied with a matrix-vector product at run time.

Even if the coupler is generally not the most critical element with respect to the overall model throughput, performance scaling was analyzed for up to 10,000 processors for different coupler kernels on different platforms; they showed standard and satisfactory behaviors given their main characterization (computation intensive, memory intensive, communication dominated).

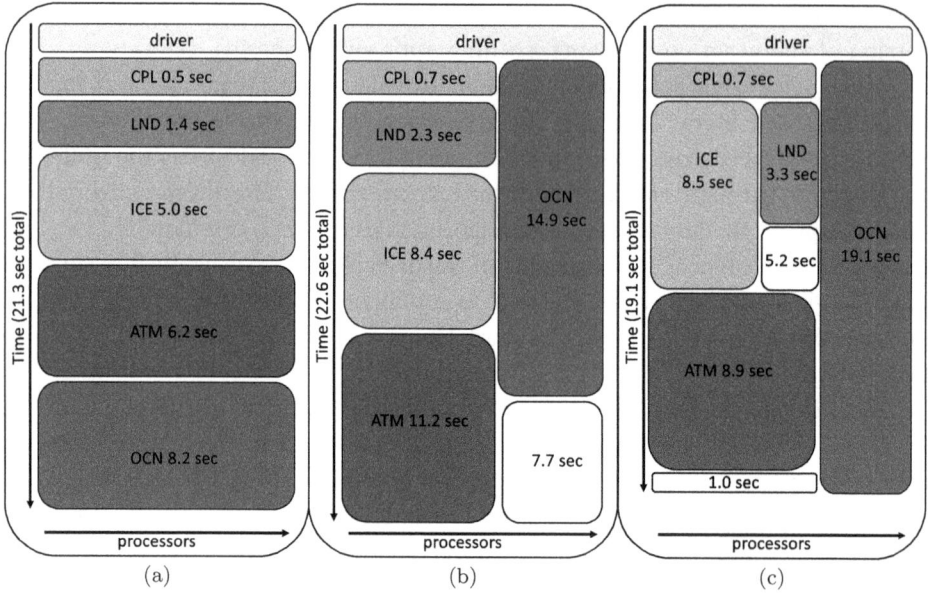

Figure 8.6.   Some of the component layouts supported by CPL7 illustrating the flexibility of the driver (numbers are fictive and provided just for illustration). In (a) all components including the flux coupler CPL are run sequentially. i.e., one after the other, on all available processors; in that case, the total run time is 21.3 seconds; in (b) hybrid layout, the ocean OCN is run in parallel with the other components, which themselves run sequentially on a subset of the available processors; the total run time is 22.6 seconds and there is some load imbalance as OCN waits for 7.7 seconds. (c) illustrates a hybrid layout similar to (b) except that the sea-ice ICE and land LND are run concurrently before the atmosphere ATM and the number of processors for OCN is reduced; the load imbalance is reduced as LND waits for 5.2 seconds, but on a relatively small number of processors, and ATM waits for 1.0 second; this layout presents the smallest runtime with 19.1 seconds in total. (Figure created by authors based on personal communication with Anthony Craig and Mariana Vertenstein from NCAR.)

In addition to NCAR in CCSM4 and CESM2, CPL7 is used at the CMCC (Centro Euro-Mediterraneo sui Cambiamenti Climatici) in Italy. Their Earth System model, CMCC-ESM2, is based on NCAR CESM in which the ocean model POP is replaced by NEMO (see Sec. 6.2.1) while keeping the same coupling infrastructure.

8.2.6   YAC

YAC (Yet Another Coupler) is a coupling library developed by the Deutsches Klimarechenzentrum (DKRZ) and the Max-Planck-Institut für Meteorologie (MPI-M) in Germany. The primary target was the ICOsahedral Nonhydrostatic (ICON) general circulation model [Wan et al., 2013] and other components coupled to ICON. YAC needs to be linked to the component codes at compilation. At run time, it performs parallel coupling exchanges between the components, ensures their transformation from the source to the target model grid and also proposes temporal transformations like accumulation or averaging.

This coupler uses its own methods for two-dimensional neighborhood search, coupling field regridding and communication. The coupling fields can be defined on regular or irregular grids on the sphere without any *a priori* assumptions on the grid structure. Special care was taken during the developments to produce parallel highly-efficient code for regridding weight calculation, while keeping an API that minimizes the interference in the component codes.

YAC is coded in C, while Fortran interfaces for all API routines are also provided. Different best practices in software development were also adopted, such as the use of an extensive unit test suite, use of Doxygen for automatic production of code documentation from annotated code sources, and source version control with Git (https://git-scm.com).

Multiple remapping methods are available in YAC, e.g., fixed-value, linear, distance-weighted, distance Gaussian weighted, first and second order conservative, hybrid cubic Bernstein-Bézier patch. Pre-defined weights can also be read from files. YAC also provides the option to use fallback regridding methods in case one method fails for some points. For example, a conservative remapping may be followed by first-order polynomial extrapolation for target cells that would have no value with the conservative interpolation because of non-matching sea-land masks.

During the initialization phase, a global search is performed once for each pair of source and target grids for which any regridding is required. A basic intersection computation using a tree-based algorithm is done between all cells of both grids using their bounding circles. The result is a prerequisite for all regriddings. If this first computation yields to a possible intersection between the cells, then a regridding-specific search is performed by the source processes, each one treating the target cells assigned to it in the initial global search. For distributed grids, the description of the grid on each process needs to include a halo with a width of at least one grid cell, and the parallel search will involve cells on the neighbor processes when needed.

YAC can explicitly consider latitude and longitude edges in addition to great-circle edges; this can make a significant difference for identification of the intersection between the cell edges involved in the conservative remapping, especially for cells of regular grids close to the poles. For the coupling field exchanges, YAC uses MPI. The behavior of the put and get is very similar to OASIS3-MCT. The component processes communicate directly with other component processes depending on the coupling configuration. For sending coupling fields, a non-blocking buffered send is called by each source process, i.e., it will return as soon as the coupling field is stored in an internally-managed buffer. On the target side, a standard blocking operation is implemented for receiving the coupling field, i.e., the call will not return before the data is effectively received. Specific tests have shown reasonable performance of YAC initialization phase and good scaling of the coupling communication involving the remapping.

YAC is available to the climate modeling community. In the next-generation of MPI-M ESM, YAC couples the ICON ocean and atmosphere components. DKRZ also aims to make YAC a part of the German national ESM strategy (NatESM). Some algorithms of the coupler, in particular the sophisticated conservative remapping algorithm are also reused within other software projects like the Climate Data Operators (CDOs, see https://code.mpimet.mpg.de/pr ojects/cdo). Future plans include built-in support for the coupling of vector fields, support for changing land-sea masks at runtime, and possibly an OpenMP thread-level parallelization of the initialization phase.

8.2.7   C-Coupler2

The development of the Community Coupler (C-Coupler) family started in 2010 in China in order to provide extensive coupling functions to a strongly increasing number of model users. In a manner similar to OASIS3-MCT and YAC, C-Coupler2 works as a library which API routines are called in the component codes, without a driving layer. C-Coupler2 can exchange coupling fields between components running as separate executables or within one executable on non-overlapping, partially overlapping, or overlapping sets of processes. For the regridding, the C-Coupler2 can use weights read from an existing file or can generate the weights with its own parallel library, CoR1. This library supports 2D horizontal regridding, possibly combined with a time-evolving 1D vertical interpolation for grids with sigma or hybrid vertical coordinate (but without any parallel decomposition along the vertical dimension).

Coupling exchanges are determined through the combination of API calls in the codes and a set of XML-formatted configuration files. The API contains calls for expressing the horizontal grid parallel decomposition thanks to global grid cell indexes (as in OASIS3-MCT and MCT), for sending or receiving the coupling fields using MPI, and for explicit field regridding. In particular, the coupling procedure may perform some data type transformation, and apply time operations such as lag, interpolation or averaging, as chosen by the user.

The C-Coupler2 includes facilitations for model nesting supporting the coupling between multiple copies of the same executable, each one having its own input parameters and data and being registered as a separate component. It also supports incremental coupling in the sense that an existing multi-component application internally using any coupling software can be considered as one component and coupled to other components. Finally, C-Coupler2 guarantees exact restart of coupled applications and provides some logging and debugging capability.

The reliability of the C-Coupler2 developments is regularly tested through a series of test cases. C-Coupler2 is used in more than 10 models at several institutions in China, for example the National Meteorological Center, National Marine Environmental Forecasting Center, National Climate Centre, First Institute of Oceanography, Institute of Atmospheric Physics, and Tsinghua University.

8.2.8   *MOAB-TempestRemap*

MOAB-TempestRemap is a relatively new coupling software based on the Mesh Oriented database library [MOAB, Tautges et al., 2004] and the TempestRemap package [Ullrich et al., 2016; Ullrich and Taylor, 2015]. MOAB-TempestRemap is developed in the framework of the Energy Exascale Earth System Model [E3SM, Golaz et al., 2019, https://e3sm.org], an ongoing, state-of-the-art Earth system modeling project funded by the DOE in the US.

The E3SM coupling infrastructure originally used CPL7 and followed its hub-and-spoke model with the coupling field regridding using weights pre-generated offline. E3SM moved to MOAB-TempestRemap distributed coupling approach so to allow recalculation of the regridding weights during the run.

MOAB is a distributed mesh data structure that stores the list of meshes only local to the process which leads to a high degree of memory compression. The connectivity and adjacency information are also stored with the local meshes. Its fully mesh-aware implementation allows MOAB-TempestRemap to efficiently recalculate the regridding weights dynamically during the run. TempestRemap focuses on the mathematically rigorous implementation of regridding algorithms and generates consistent regridding operators for spectral element discretization without the need to go through an intermediate dual mesh. Like ESMF and YAC, TempestRemap identifies, for each target point, the source cell of its location by querying a tree data structure of the source elements and the corresponding element is then marked as a contributor to the regridding computation. Then the intersection of the source and target grid meshes are computed using an advancing front algorithm that uses the mesh face adjacency information available in the MOAB data structure.

MOAB-TempestRemap specificity is its relatively low-memory needs and its support of moving or adaptive grids, without the need for dual grid and therefore preserving high-order spectral accuracy.

8.3   Analysis of coupler or coupling software

Although their implementations differ substantially, all couplers or coupling software used in Earth System modeling, and typically for atmosphere-ocean coupling, offer basic coupling functions. These basic functions are managing the data transfer between the component models in a coordinated way as well as transforming this data from the source to the target component grid. Many coupling infrastructures also provide additional utilities, not directly related to coupling, such as calendar management, logging, or error handling. Basic coupling functions are detailed in Sec. 8.3.1.

A closer look leads also to the conclusion that the current coupling software can be classified into two main broad categories. In the "external coupler or coupling library" approach, the component models typically remain separate executables and a coupling library linked to the components ensures the exchanges and the regridding of the coupling fields. In the "integrated coupling framework" strategy, the original component models are split into elementary units and a single coupled application integrating these units is rebuilt using the coupling framework software. The characteristics of each category of coupling software are presented in Sec. 8.3.2.

8.3.1   *Basic coupling functions*

8.3.1.1   *Data transfer and coordinated execution of components*

The most basic functionality of a coupling software is to ensure the communication of coupling data between the component codes in a coordinated way. The different coupling software may implement different schemes of communication.

On one hand, when model components run concurrently as separate executables on different processes, exchange of coupling fields is typically done in their timestep loop via send and receive calls to the coupling library. This is illustrated in Fig. 8.7a). Below these calls, the transfer of data is done via message passing or other protocols from the source component to the target component, either directly or through additional coupling processes. The Message Passing Interface (MPI) is the communication protocol mostly used today. Other techniques have also been used in the past, such as PVM (Parallel Virtual Machine), TCPIP (Transmission Control Protocol/Internet Protocol), SVIPC (System V Inter Process Communication), or CORBA (Common Object Request Broker Architecture), but these are much less popular today. In this first case, there is no explicit coordination of the components by the coupling software. Their synchronization relies on the coupling exchanges and it is the user's responsibility to implement a suitable coupling algorithm. The synchronization of the components is ensured by the blocking receive instructions, i.e., the component is not able to advance in time if it does not receive needed inputs.

On the other hand, when the component models are executed sequentially, i.e., one after the other, on the same set of processes, the coupling fields can be exchanged as arguments of subroutine calls in a shared memory context, as illustrated in Fig. 8.7(b)). In that case, a top-level driving layer explicitly

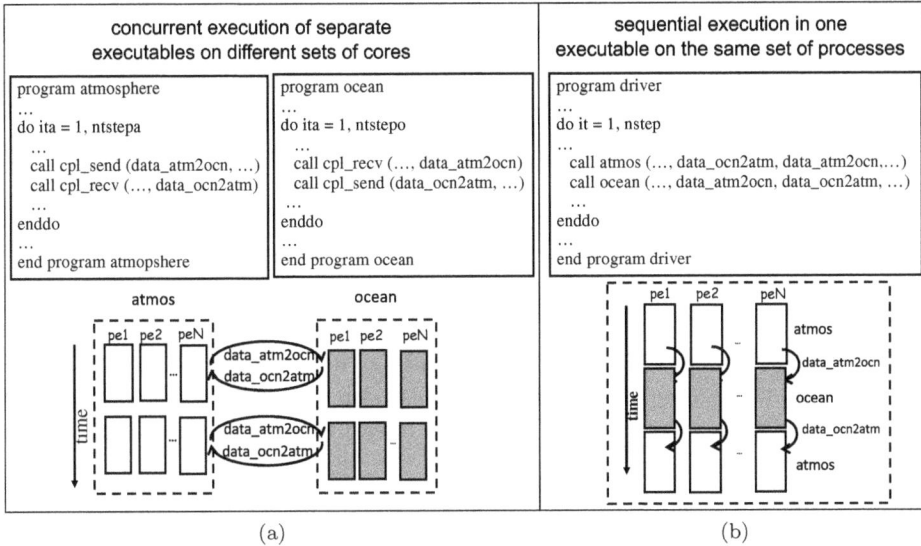

concurrent execution of separate executables on different sets of cores		sequential execution in one executable on the same set of processes

```
program atmosphere

do ita = 1, ntstepa
  ...
  call cpl_send (data_atm2ocn, ...)
  call cpl_recv (..., data_ocn2atm)
  ...
enddo
...
end program atmopshere
```

```
program ocean

do ita = 1, ntstepo
  ...
  call cpl_recv (..., data_atm2ocn)
  call cpl_send (data_ocn2atm, ...)
  ...
enddo
...
end program ocean
```

```
program driver

do it = 1, nstep
  ...
  call atmos (..., data_ocn2atm, data_atm2ocn,....)
  call ocean (..., data_atm2ocn, data_ocn2atm, ...)
  ...
enddo
...
end program driver
```

Figure 8.7. Typical coupled configurations: (a) Component models remain separate executables. They run concurrently and exchange coupling fields via calls implemented in their timestep loop. (b) Components are executed sequentially on the same set of processes and they exchange their coupling fields as argument of subroutine calls.

controls the execution of the components. This driving layer can be predefined or it can be customarily built by the user for each coupled system.

In practice, coupled applications often implement a mix of the two communication schemes between their different components.

8.3.1.2 Coupling field regridding and transformation

A majority of couplers offer regridding functionalities and also quite often other field operations such as temporal and arithmetical transformations, global conservation, field combination, etc. The following paragraphs describes these transformations.

Temporal transformations. The coupling field is provided by the source component model at a specific frequency, typically the time step frequency, which is usually shorter than the coupling frequency. Operations such as time accumulation, averaging, interpolation, minimum or maximum can be performed on the field.

Arithmetical transformation. Different arithmetical operations, such as addition or multiplication by a scalar, can be applied on the fields.

Spatial regridding. A coupling field sent by a source component model needs to be transformed for use by a target component model on its grid. This spatial transformation is called "regridding". The first step is to determine the addresses and weights that describes which and how points of the source grid contribute to the calculation of the points in the destination grid. The second step is the regridding *per se*, i.e., the multiplication of the source grid values by the regridding weights to produce values on the target grid. The most typical algorithms in the different coupling software to define regridding addresses and weights are,

(1) n-nearest-neighbor, using the values of the n closest grid points on the source grid with weights inversely proportional to their great circle distance from the target grid point.

(2) Bilinear, using the values at the 4 enclosing source grid points.

(3) Bicubic, using the values and derivatives of the coupling field at the 4 enclosing source grid points. It is generally used when it is important to conserve higher order properties of the coupling field, such as the curl of the wind.

(4) Patch-recovery, constructing multiple higher-order polynomial patches to represent the data in the source cells and average of the values at the patches. This algorithm typically results in better approximations of the target values and their derivatives when compared to the bilinear interpolation.

(5) First or second-order 2-D conservative remapping, in which the value at each target cell is computed as a weighted sum of source cell values with the contribution of a source cell being proportional to the fraction of the target cell intersected by the source cell (first order), or including additional terms proportional to the gradient of the source field in the longitudinal and latitudinal directions (second order). Conservative methods should be applied when it is important to conserve the area-integrated value of the coupling field, such as the energy flux or precipitation. For ocean-atmosphere coupling, special care has to be taken because of the sea-land mask issue (see Sec. 9.4.2.2). A second order conservative remapping ensures that a subgrid variability is reconstructed and that different target cells entirely located under the same source cell receive different values.

(6) Vertical interpolation, in which regridding of 3D coupling fields are regridded in the vertical direction. This operation — currently available in a few couplers — can be more or less complex depending on the type of vertical coordinates. If the vertical coordinate systems of the source and target models are fixed in space, as is the case with the Z coordinate, the regridding weights mapping the vertical levels of the source grid on the vertical levels of the target grid can be calculated once and used during the whole simulation. But if the position of the vertical levels of at least one grid evolves with time, which is typically the case for sigma or hybrid sigma-pressure coordinate systems, the regridding weights have to be

recalculated for each coupling exchange. This implies transferring additional fields from the model to the coupler. For example, a regridding involving a hybrid sigma-pressure vertical coordinate system requires transferring the surface pressure from the atmospheric model to the coupler at each coupling timestep, so to recalculate the vertical position of the levels each time.

Support for vector fields. Few couplers specifically support the regridding of vector fields, for which special care has to be taken. One plausible method is to project the vector field in a fixed Cartesian coordinate system and to send and treat the three components of the projected vector as separate coupling fields. The three regridded Cartesian components are then recombined and projected on the target coordinate system.

Global conservation. This can be a useful option when the regridding method is not conservative by design. To force global conservation, integrals of the coupling field on the source grid and on the target grid are calculated. The difference is then distributed on the target grid, either applying a uniform additive term, or as a multiplicative term. The latter technique has the advantage of not to alter the sign of the field but may cause problems for specific fields, for example when the area-averaged field value tends to zero.

Merging. Some couplers offer the possibility to combine different coupling fields coming from different source components. This functionality is of interest when the expected target field is the combination of different source fields. For example, net water flux into the ocean is the difference between precipitation and evaporation. Merging may include the possibility of weighting by fractions of different surface types, as discussed in more details in Sec. 9.4.2.3.

Flux computation. In some couplers, like in CPL7, more complex computation of physical quantities such as surface fluxes based on oceanic surface fields and atmospheric first-level quantities are proposed. In those cases, the coupler qualitatively also becomes a scientific module of the coupled system.

8.3.2 *Advantages and disadvantages of the different coupler implementations*

As noted above, the current coupling software can be broadly classified into two main categories: the "external coupler or coupling library", and the "integrated coupling framework" (see Fig. 8.8). In the first approach, the component models remain separate executables and the original codes are modified as little as possible. The code of the component models is instrumented with calls to the coupling library API and the synchronization of the components is implicitly ensured by the algorithm of the coupling exchanges. The communication, regridding and other transformations on the coupling fields are done either

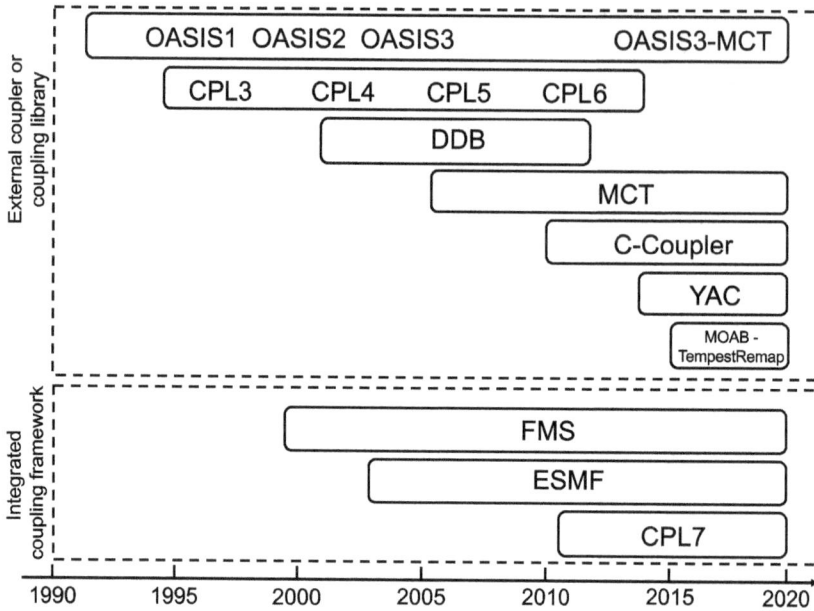

Figure 8.8.   Classification of the generic coupling software described in Sec. 8.2 into two main categories, the "external coupler or coupling library" and the "integrated coupling framework".

directly by the coupling library, as in the C-Coupler2, DDB, MCT, MOAB-TempestRemap, OASIS3-MCT (see Fig. 8.1(d)) and YAC, or via a central coupling hub that can be a monoprocess, as in CPL5, OASIS1 or OASIS2 (see Fig. 8.1(a) and (b)) or parallel, as in CPL6 or OASIS3 (see Fig. 8.1(c)).

The "integrated coupling framework" approach involves splitting the original component codes into initialize, run and finalize units, adapting them to standard data structures and routine interfaces, and rebuilding a single integrated application based on these units. In this approach, a driving layer explicitly calls the different component and coupling units and controls their execution. This driving layer can be pre-defined, as is the case for CPL7, FMS and NUOPC, or user-built for each coupled application, as for ESMF when the NUOPC layer is not used.

The main advantage of the "external coupler or coupling library" approach is the low degree of intrusiveness in the components. These keep their original structure and can easily be maintained and used as stand-alone executables in their original field of application, even when instrumented with the coupling library calls. Further, components can keep their own compiling environment and the risk of conflict, for example in terms of namespace or I/O, is minimized. In view of these merits, the external coupler approach has been generally chosen by communities using models developed by different research groups and where imposing coding rules is not practical. The OASIS couplers have succeeded in this regard across groups in Europe and beyond.

Couplers belonging to the "integrated coupling framework" category force the adoption of the initialize-run-finalize code structure, standard data structures and calling interfaces, and are therefore more intrusive. This approach supposes agreement among modeling groups on using the same framework and following the same coding rules. Such an agreement may not be easy to reach. However, the approach forces developers to define explicit interfaces for each scientific module and therefore favors code modularity, which is an important best-practice to promote.

One advantage of the "external coupler or coupling library" approach is that different coupling algorithms can be easily realized as the send and receive calls can be flexibly placed anywhere in the component model timestep. From that point of view, the integrated approach is more limited as the coupling exchanges necessarily happen at the end or at the beginning of the component run unit. Although this can simplify program flow, it can affect the time sequencing and may require splitting the run unit into smaller sub-units in order to respect the original scientific formulation.

One drawback of the "external coupler or coupling library" approach is a possible lack of efficiency. Multiple-executable systems necessarily imply that the components are running concurrently on separate processors, since operating systems do not allow different executables to share the same hardware processors. The coupling exchanges therefore occur between different processes via message passing or other protocols. The speed of the transfer depends on the message passing latency and on the bandwidth available to transmit the information. The forced component concurrency can also lead to a waste of resources if one component necessarily needs to wait until another completes its coupling period. In that case, the coupling might be more efficient if the components were run sequentially on the same set of resources within one same executable.

In contrast, the "integrated coupling framework" offers more opportunities for optimization. First, components may possibly share the coupling fields via their memory, perhaps using pointers, which reduces their duplication and the communication time. Also, the driving layer may allow the different components to be flexibly placed on arbitrary subsets of computing resources in a sequential, concurrent, or mixed mode (as shown in Fig. 8.6). This should help to define an overall sequencing of the components that optimizes the coupled model performance.

8.4 Perspectives

This chapter has highlighted the concept that coupling software offers different modeling groups that specialize in the development of specific climate model components a way to build larger applications by incorporating other components developed elsewhere through efficient interfaces. Coupling technologies are therefore very helpful as no individual group can today aim to develop from scratch and master all aspects of very complex systems, such as climate models.

The technical interoperability between the components provided by the adoption of the same coupling technology does not ensure their scientific interoperability. Even if the components agree on how to exchange information, they also have to agree on what information to exchange. This is why the development of coupling metadata (i.e., data providing information about coupling data) describing the component model interface, the coupling fields they expect, and the ones they can provide, is gaining growing interest in the community although no widely-adopted standard exists yet.

We have analyzed the advantages and drawbacks of the two approaches to coupling software, the "external coupler or coupling library" and the "integrated coupling framework". These two approaches answer different needs emerging from different modeling communities with different strengths and constraints, and therefore both are likely be maintained and further developed in the short to medium term future. Of course, this categorization is somewhat simplistic and, in many cases, a software classified into one category presents characteristics from the other category. For example, the latest versions of OASIS3-MCT, which we classified as an "external coupling library", now support coupling exchanges between components deployed sequentially within one same executable, on the same or overlapping sets of processes. Another example is the ability introduced in ESMF, classified as an "integrated coupling framework", to support heterogeneous coupling between ESMF components implemented as web services.

Over the past three decades, several research groups have developed different coupling software relatively independently and today different scientific communities benefit from different solutions adapted to their needs. However, there is also basic agreement that sharing more software would be beneficial as it reduces the duplication of efforts, favors transfer of knowledge and scientific interactions. The merge of MCT into OASIS is an example of such successful interaction. Moving forward, the potential benefit of much closer collaboration should also be evaluated even if the different design approaches, answering specific needs of different communities, are likely to persist for some time in the future.

Further improvement in coupling software performance will be more challenging. Most gains in the last decade came from faster hardware on a per-processor basis, parallelization, and improvement in communication algorithms. Future-generation computers are likely to consist of orders of magnitude more processors, slower, heterogeneous, and with less and slower memory. Moving into the exascale era requires, for coupling technology as for other software, finding additional opportunities for parallelism, increased concurrency, and improved communication mechanisms to better overlap communication with computation. Given the uncertainty of future hardware, fast adaptation will certainly be a competitive advantage.

Chapter 9

Coupling Algorithms and Specific Coupling Features in CGCMs

9.1 Introduction

In Ch. 8 we made a detailed presentation of technical methodologies in the implementation of data exchanges between components of Coupled General Circulation Models (CGCMs, see also Ch. 7). In this chapter, we describe how specific coupling algorithms are implemented in a subset of active CGCMs. For historical and practical reasons, present-day physical interfaces in these models are very often the result of an ad-hoc approach motivated by special features of the component models in their stand-alone configurations. In general, however, coupling algorithms used by different modeling groups have been carefully implemented to respect physical constraints such as conservation laws. We also present in this chapter a proposition from the European PRISM project for a revised physical interface between the main components of CGCMs, i.e., the atmosphere, ocean, sea-ice and land. This interface is currently being implemented at least partially in several models. Finally, we analyze how specific coupling features, such as the component sequencing or the flux calculation, are implemented in CGCMs and discuss how these implementations can generate inconsistencies in time or in space.

9.2 Coupling algorithms in selected CGCMs

9.2.1 *Implementation in European CGCMs*

The coupling algorithms of several CGCMs developed in Europe have many similarities. A schematic of the coupling exchanges is shown in Fig. 9.1. In the CGCMs we are considering in this section, the ocean includes a sea-ice model and the atmosphere a land scheme. At the end of each coupling period, the ocean-ice model sends surface variables to the atmosphere-land model, which uses them as boundary conditions during the next coupling period. If the subscripts "o" and "i" denote values over water and ice, respectively, the variables sent by the ocean include surface temperature (T_o, T_i), albedo (Alb_o, Alb_i),

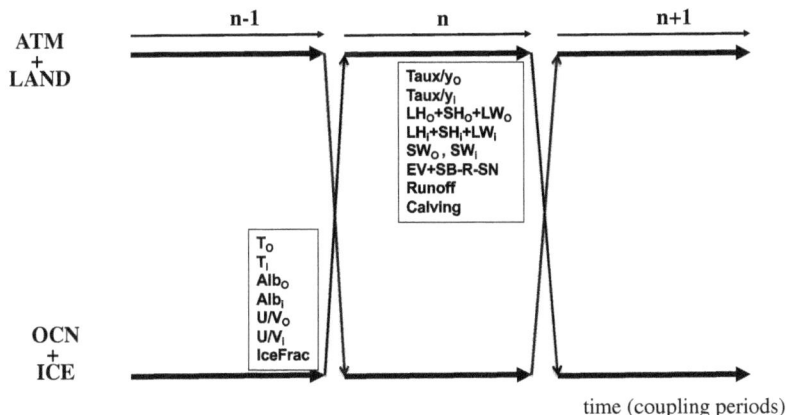

Figure 9.1.   Typical ocean-atmosphere asynchronous coupling algorithm implemented in many European CGCMs. The horizontal arrays represent the time evolution during three coupling periods, at the bottom for the ocean-ice model and at the top for the atmosphere-land model. Details of the coupling exchanges are provided in the text.

velocities in the longitudinal (U_o, U_i) and latitudinal directions (V_o, V_i), and ice fraction (IceFrac). In return, the atmosphere-land model sends surface fluxes of momentum, energy and water averaged over the coupling period in order to ensure global conservation of integrated quantities. The fields sent by the atmosphere include wind stress in the longitudinal (T_{aux_o}, T_{aux_i}) and latitudinal (T_{auy_o}, T_{auy_i}) directions, sensible heat (SH_o, SH_i) and latent heat (LH_o, LH_i) fluxes, and longwave (LW_o, LW_i) and shortwave (SW_o, SW_i) fluxes. The water fluxes are also passed from the atmosphere-land to the ocean-ice, i.e., evaporation (EV), sublimation (SB), rain (R), and snow (SN), as well as runoff of rainfall (Runoff) and the breaking of ice from glaciers to the ocean (Calving).

This algorithm has been implemented in many European CGCMs used in CMIP6 (see https://portal.enes.org/models/earthsystem-models), with some special features.

- In CNRM-CM6-1, developed by CNRM-CERFACS (Centre National de Recherches Météorologiques — Centre Européen de Recherche et de Formation Avancée en Calcul Scientifique), the algorithm includes a separate river routing model.
- In IPSL–CM6, assembled by IPSL (Institut Pierre Simon Laplace), the algorithm takes care of the possible sea-land mask inconsistencies between the atmosphere and ocean models (see Sec. 9.4.2.2).
- In EC-Earth3, developed by a Europe-wide consortium of 27 research institutes from 10 European countries, the coupling includes a runoff mapper running as a separate executable.

- In MPI-ESM, the Earth System Model developed by MPI-M, exchanges among component models include carbon fluxes.
- In HadGEM3-GC31, set up by the UK MetOffice, 5 ice categories are considered for sea-ice surface temperature, ice fraction, sea-ice depth, snow depth, and sublimation.

The coupling algorithm we have been describing is called asynchronous because component models run concurrently and the coupling fields exchanged at the end of a coupling period are used in the target model as boundary condition for the next coupling period. Coupling periods typically lie in the range of one to three hours. Also, surface variables are mapped from the ocean-ice to the atmosphere-land while fluxes are calculated in the atmosphere-land model, although this is not fully physically justified as we will discuss in Sec. 9.4.2.1.

9.2.2 ECMWF-IFS coupling algorithm

The ocean-atmosphere coupling algorithm implemented in the CGCM developed at the European Centre for Medium-Range Weather Forecast (ECMWF) is quite different from the one discussed in the previous subsection. In particular, it involves a wave model and the three component models run sequentially as illustrated in Fig. 9.2.

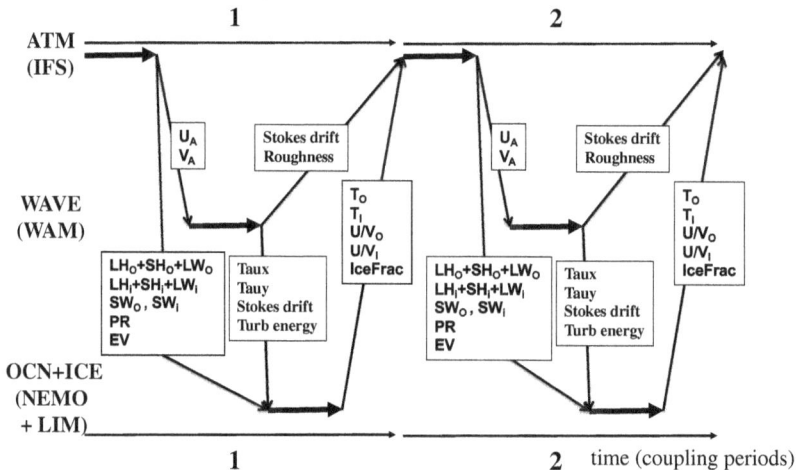

Figure 9.2.   Coupling algorithm implemented in ECMWF-IFS involving three component models running sequentially one after the other. The horizontal arrays represent the time evolution during 2 coupling periods, at the bottom for the ocean model (NEMO) including a sea-ice component (LIM), in the middle for the wave model (WAM), and at the top for the atmosphere (IFS). Details of the coupling exchanges are provided in the text.

In ECMWF-IFS, no part of the code is specifically labeled as "coupler". The atmosphere model IFS (Integrated Forecast System) acts as the master model calling sequentially the wave model WAM (WAve Modeling) and ocean model NEMO (including the LIM ice model) as subroutines as well as the regridding routines. The atmosphere transfers to the wave subroutine surface wind velocities (U_A, V_A) as arguments. The wave model produces the wind stress in the longitudinal and latitudinal directions (T_{aux}, T_{auy}) and the turbulent kinetic energy associated with wave breaking (Turb energy) as well as the Stokes drift, which are transferred to the ocean. The Stokes drift together with surface roughness (Roughness) are sent to the atmosphere to be used in the momentum and heat flux calculations. Heat and water fluxes calculated by the atmosphere are transferred to the ocean subroutine, which includes a sea-ice calculation. The ocean-ice then transfers back to the atmosphere surface properties such as temperatures (T_o, T_i) and longitudinal and latitudinal velocities $(U/V_o, U/V_i)$ over water and ice, and ice fraction (IceFrac).

The coupling periods between the atmosphere and wave and ocean models can be different. Typically, the atmosphere calls the wave model at each time step, which is 1,800 seconds for the current low-resolution version of ECMWF-IFS and 1,200 seconds for its high-resolution version, while it calls the ocean every hour, so 2 or 3 times less frequently; this is justified by the different time scales of the processes involved.

9.2.3   *RPN coupling algorithm*

The CGCM developed by RPN (Centre de Recherche en Prévision Numérique) from the Canadian meteorological and climatic services (Environment and Climate Change, Canada) couples the Global Environmental Multiscale (GEM) atmospheric model to the NEMO ocean model [Smith et al., 2018] using the algorithm illustrated in Fig. 9.3.

At the end of its first coupling period, the ocean-ice model sends to the atmosphere the following surface fields over water and ice: temperature $(T_{o/i})$, sensible and latent heat fluxes $(SH_{o/i}, LH_{o/i})$, wind stress in both directions $(T_{aux_{o/i}}, T_{auy_{o/i}})$, and ice fraction (IceFrac). (Z_M and Z_H used in the calculations are the height of the atmospheric first level wind and temperature). The atmospheric model then runs for 2 coupling periods and provides the shortwave (SW_A) and longwave (LW_A) fluxes, solid and liquid precipitation (PR), and state variables at the first level in the atmosphere (temperature T_A, velocities U_A and V_A, humidity q_A, and pressures P_A and P_O) to the ocean. The ocean can then start its second coupling period while the atmosphere runs for its third coupling period still using fields provided by the ocean at the end of its first coupling period. Afterwards, the coupling keeps this lagged asynchronicity as illustrated in Fig. 9.3.

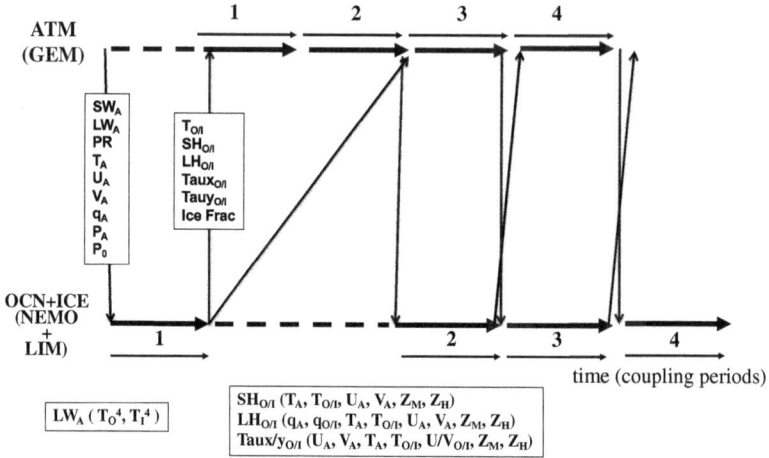

Figure 9.3. Coupling algorithm implemented in the CGCM developed RPN. The horizontal arrays represent the time evolution during 4 coupling periods, at the bottom for the ocean model (NEMO) including the LIM sea-ice model and at the top for the atmosphere model, GEM. Details of the coupling exchanges are provided in the text.

9.2.4 *CESM2 and CMCC-CM2 coupling algorithm*

Figure 9.4 is a sketch of the coupling exchanges implemented in CESM2 and CMCC-CM2, developed respectively at NCAR and CMCC (Centre Euro-Mediterraneo sui Cambiamenti Climatici). The participating models are CAM for the atmosphere (Atm), CLM for the land (Lnd), CICE for the sea-ice (Ice) and POP -in CESM2- or NEMO -in CMCC-CM2- for the ocean (Ocn). Exchanges also include a "flux coupler" (Cpl), which calculates the surface fluxes based on variables sent by the ocean and the atmosphere. Cpl also contains the "accumulate-to-ocean" module (acc2ocn), which receives fluxes calculated by the other models, accumulates them, and sends them to the ocean. Finally, a "merge" module aggregates the surface fluxes calculated in Ice, Lnd and Cpl and sends them to Atm. Figure 9.4 does not represent fields exchanged with the runoff and glacier models.

The exchanges among the different models proceed as follows:

- From Lnd, Ice and Cpl to Atm (via the "merge" module): upward longwave (LW_u), latent (LH) and sensible (SH) heat fluxes, wind stress (T_{aux}, T_{auy}), evaporation (EV), and albedo (Alb);
- From Lnd and Ice to Atm: surface temperature (Tl, Ti), snow depth (SNd), ice fraction (IceFrac, for Ice) and some boundary layer parameters like the wind at 10m (W10m) and the temperature at 2m ($T2m$);
- From Atm to Lnd, Ice, and Cpl: surface winds (Ws), temperature (Ts), pressure (Ps), humidity (qs);

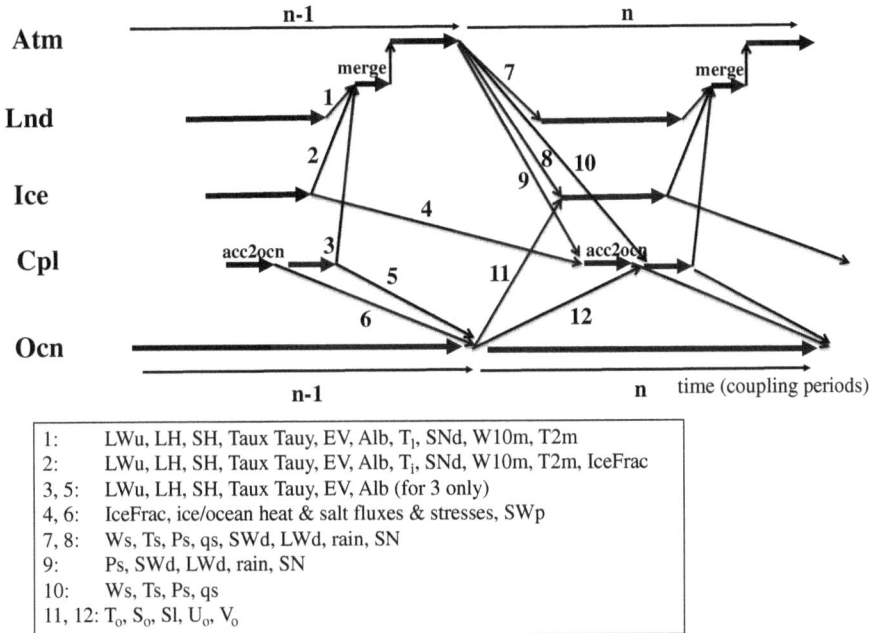

1:	LWu, LH, SH, Taux Tauy, EV, Alb, T_l, SNd, W10m, T2m
2:	LWu, LH, SH, Taux Tauy, EV, Alb, T_i, SNd, W10m, T2m, IceFrac
3, 5:	LWu, LH, SH, Taux Tauy, EV, Alb (for 3 only)
4, 6:	IceFrac, ice/ocean heat & salt fluxes & stresses, SWp
7, 8:	Ws, Ts, Ps, qs, SWd, LWd, rain, SN
9:	Ps, SWd, LWd, rain, SN
10:	Ws, Ts, Ps, qs
11, 12:	T_o, S_o, Sl, U_o, V_o

Figure 9.4.   Coupling algorithm implemented in CESM2 and CMCC-CM2. The horizontal arrays represent the time evolution during 2 coupling periods, at the bottom for the ocean model (Ocn), above it for the flux coupler (Cpl), in the middle for the sea-ice model (Ice), above it for the land model (Lnd), and at the top for the atmosphere model (Atm). Details of the coupling exchanges are provided in the text.

- From Atm to Lnd, Ice, and Ocn via acc2ocn : surface pressure (Ps), shortwave and longwave downward fluxes (SWd, LWd), rain, and snow (SN);
- From Ice to Ocn via acc2ocn: ice fraction (IceFrac), ice/ocean heat & salt fluxes, ice/ocean stresses and the shortwave penetration (SWp);
- From Cpl to Ocn: upward longwave (LWu), latent (LH) and sensible (SH) heat fluxes, evaporation (EV), and wind stress (Taux, Tauy);
- From Ocn to Cpl and Ice: surface temperature (T_O), salinity (S_O), velocity (U_O, V_O), sea surface slope (Sl).

In this algorithm, the turbulent diffusive fluxes (latent, sensible), wind stress, evaporation, albedo and upward longwave are calculated at resolution of the surface in the land, ice and flux coupler modules for the atmosphere/land, atmosphere/ice and atmosphere/ocean parts, respectively. These fluxes are aggregated together by the "merge" module and sent to the atmosphere, which uses them for the same period. For the free ocean, the turbulent diffusive fluxes are calculated by the flux coupler. When ice is present at the ocean surface, the sea-ice model calculates the ice/ocean heat and salt fluxes as well as the penetrative shortwave and sends these fields to the ocean via the acc2ocn module. The shortwave and longwave downward fluxes, and water fluxes (rain and snow) are calculated by the atmosphere and sent to the ocean also via the acc2ocn module.

The land and ice surface fluxes are computed in the land and ice models while ocean surface fluxes are computed separately in the flux coupler (Cpl); this allows for more frequent coupling for the fast land and ice surface processes, and less frequent coupling for the slower ocean processes (not included in Fig. 9.4).

9.3   PRISM revised ocean-atmosphere physical coupling interface

The PRogramme for Integrated earth System Modelling (PRISM) funded by the European Commission over the 2001–2004 period was the first in a series of infrastructure projects for climate modeling in Europe. PRISM discussed a proposition for a revised physical interface linking the main components of climate models (i.e., atmosphere, ocean, sea-ice, and land surface). The resulting interface is currently being implemented, at least partially, in several models. This interface introduces two specific surface modules: (1) Surface layer turbulence (SLT), and (2) Ocean surface (OS). This added modularity allows physically consistent interpolations across grids, ensures that exchanges between model components are "process-based", and helps controlling unstable computations by distinguishing faster and slower processes.

The revised physical coupling interface is based on the following principles:

- To identify physically based interfaces across which the conservation of energy, mass and momentum can be ensured;
- To identify which process needs to be computed by which component/module and ensure that there is no duplication or inconsistency in these computations;
- To consider numerical constraints (stability, impact of different component and coupling time steps, ...) and interpolation constraints (subgrid scale heterogeneity issues, local conservation, ...).

The interface proposed by PRISM is schematically shown in Fig. 9.5. In this figure, exchanges are represented by groups of fields attached to solid arrows and numbered from (1) to (8). In the remainder of the text, each field is identified by a two-digit code, where the first digit is the exchange number and the second is the field number; for example, field 2.6 in Fig. 9.5 stands for evaporation.

The Surface layer turbulence (SLT) module receives information on the surface layer of the atmospheric model (exchange (3)) and on the surface boundary conditions from the Ocean surface (OS) module and from the Land surface model (exchange (5)). The SLT module passes the coefficients of the bulk formulas for surface turbulent flux calculation to the OS module and Land

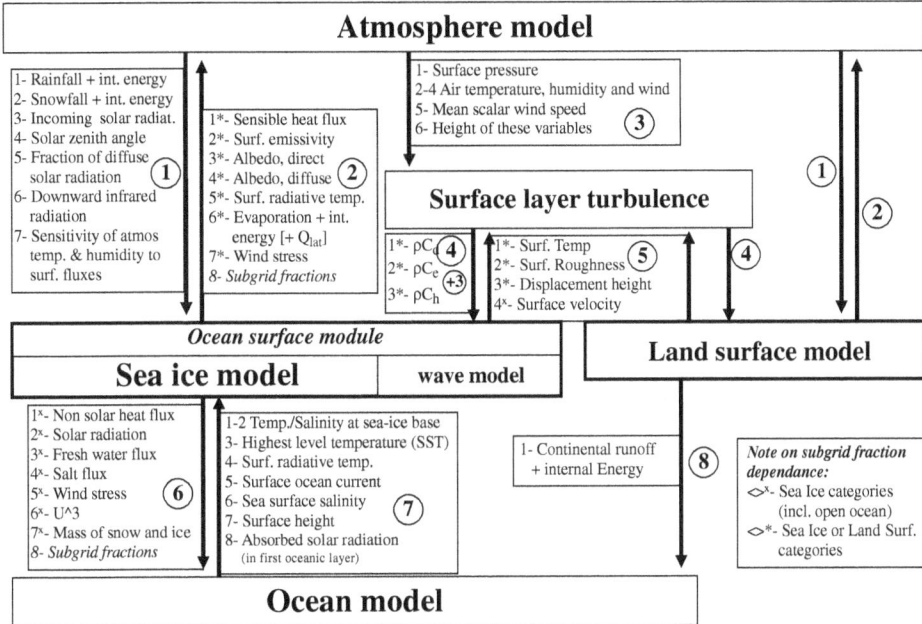

Figure 9.5. The PRISM revised physical interface proposed for coupling atmosphere, ocean, sea-ice and land surface components. The exchanges are detailed in the text. (Figure from Valcke and Guilyardi (2008). Used with permission.)

surface model (exchange (4)). The SLT runs at a resolution that is the finest among the models with which it interacts.

The OS module computes the fast ocean and sea-ice surface processes separately from the deep ocean processes, which are computed in the Ocean model. The OS module exchanges fields with the atmosphere ((1) and (2)) and the Ocean model ((6) and (7)). A wave model can be included in the OS module to provide sea surface roughness (field 5.2) to the SLT module.

9.3.1 Exchanges of energy

The OS module, via the sea-ice model, provides the Ocean model with the net solar radiation (6.2) and heat fluxes other than solar (6.1) entering the ocean surface. In return, the Ocean model provides the OS module with temperature at the sea-ice base (7.1) used by the sea-ice model to compute the oceanic heat flux at the ice-ocean interface, SST (7.3) required for the calculation of the atmospheric turbulent heat fluxes, sea surface radiative temperature (7.4) required for the calculation of the long-wave radiation over leads, surface height (7.7) further transferred to the SLT module, and the fraction of solar radiation absorbed by the first oceanic layer (7.8) needed for the computation of the energy budget of leads by the sea-ice model.

The Atmosphere model provides the OS module with the incoming solar radiation possibly for different spectral intervals (1.3), solar zenith angle (1.4), fraction of diffuse solar radiation (1.5), and downward infrared radiation (1.6). In return, the OS module provides surface emissivity (2.2), albedo for direct and diffuse radiation (2.3 and 2.4), possibly for different spectral intervals consistently with the partitioning of incoming solar radiation, and surface radiative temperature (2.5); these fields are calculated while solving the surface radiation budget either over free ocean or over sea-ice.

In order to evaluate the surface turbulent fluxes, the Atmosphere model provides the SLT and OS modules with surface pressure (3.1), air temperature (3.2), specific humidity (3.3), wind components (3.4) and mean scalar wind speed (3.5), and the height of the level at which all these variables are calculated (3.6). The Atmosphere also provides the OS module with the sensitivity of atmosphere temperature and humidity to surface fluxes (1.7). The OS module provides the surface temperature (5.1), surface roughness (5.2), displacement height (5.3), and surface velocity (5.4) to the SLT module. With fields received through exchanges (3) and (5), the SLT module can compute the exchange coefficients for sensible heat (4.2) and moisture (4.3), which are passed to the OS module that calculates the sensible heat flux (2.1) and transfers them to the Atmosphere model. Exchanges (1), (2), (3) and (4) allow for an implicit calculation of the energy fluxes over the whole column from the base of the sea-ice to the top of the atmosphere

9.3.2 *Exchanges of mass*

The Atmosphere model provides the OS module with rainfall (1.1) and snowfall (1.2), and associated internal energies, both used by the sea-ice model and the ocean model. In return, the OS module provides evaporation/sublimation (2.6), based on the exchange coefficient for moisture (4.3). This field is needed for the calculation of the hydrological cycle in the atmosphere.

The OS module transfers to the Ocean model the net freshwater flux (6.3), net salt flux (6.4) and total mass of snow and ice (6.7). The net freshwater flux (6.3) results, on one hand, from the net atmospheric water flux over open ocean (Rainfall+Snowfall−Evaporation) and on the other hand from snow melting on top of sea-ice, ice growth/melting, snow-ice formation, runoff of rainfall through sea-ice into the ocean, snowfall and rainfall over leads, and evaporation over leads. The net salt flux (6.4) is provided by the sea-ice component and results from ice growth/melting and snow-ice formation. In return, the ocean component provides salinity at the sea-ice base (7.2), which is used by the sea-ice model to compute the freezing point of sea water and the salinity of newly-formed sea-ice and snow ice, and the sea surface salinity (7.6). The total mass of snow and ice (6.7) is provided to the ocean model to compute the depression of ice below the water level. Finally, the land surface model provides the continental run-off to the ocean (8.1) and the associated internal energy.

9.3.3 *Exchanges of momentum*

The surface turbulent wind stress is computed by the OS module. The computation uses the drag coefficient (4.1) provided by the SLT module, the wind at the lowest level (3.4), its module (3.5) and its height (3.6) provided by the atmosphere via the SLT module, and the surface ocean currents (7.5) provided by the ocean. The OS module then transfers the wind stress to the atmosphere (2.7) and to the ocean (6.5). The OS module also computes the "wind work" U^3 (6.6) and provides it to the ocean.

9.3.4 *Subgrid scale computations*

The representation of subgrid scale processes is a central concern in Earth System models and must be properly considered by physical interfaces. Difficulties can arise because of different grid resolutions or multiple sea-ice or surface categories into one grid cell. Two situations can happen; the coarser grid component can either accommodate subgrid scale information (like multiple surface category "tiling" in atmosphere) or it can only handle an average value. In the latter case, the averaging should be done on the finer grid (i.e., in the OS module or Land surface component); in the former case, each subgrid flux together with the corresponding subgrid fraction should be passed from the finer to the coarser grid. Hence, fields of exchanges 2, 5 and 6 become arrays. In that case, subgrid fractions of the different sea-ice or surface categories (2.8) are passed from the OS module to the atmosphere (which will blend the information either in its boundary layer, up to a blending height or even all the way up in the atmospheric column — e.g., delocalized physics [Vintzileos & Sadourny, 1997] — depending on the complexity of the tiling scheme in the atmosphere) and subgrid fractions of different sea-ice categories including open ocean (6.8) are passed from the sea-ice model to the ocean model.

9.3.5 *Time sequence*

A schematic of a possible time sequence for ocean/atmosphere/sea ice coupling via the SLT and Ocean surface modules is presented in Fig. 9.6. This sequencing allows for the atmosphere time integration of fluxes to remain implicit.

A classification of the different components involved in the exchanges in terms of speed of processes gives, going from the slowest to the fastest: (1) the Ocean model, which includes deep ocean processes, (2) sea-ice model (excluding energy flux calculation), (3) Ocean surface module (for energy flux calculation), (4) Surface layer turbulence module, and (5) Atmosphere model.

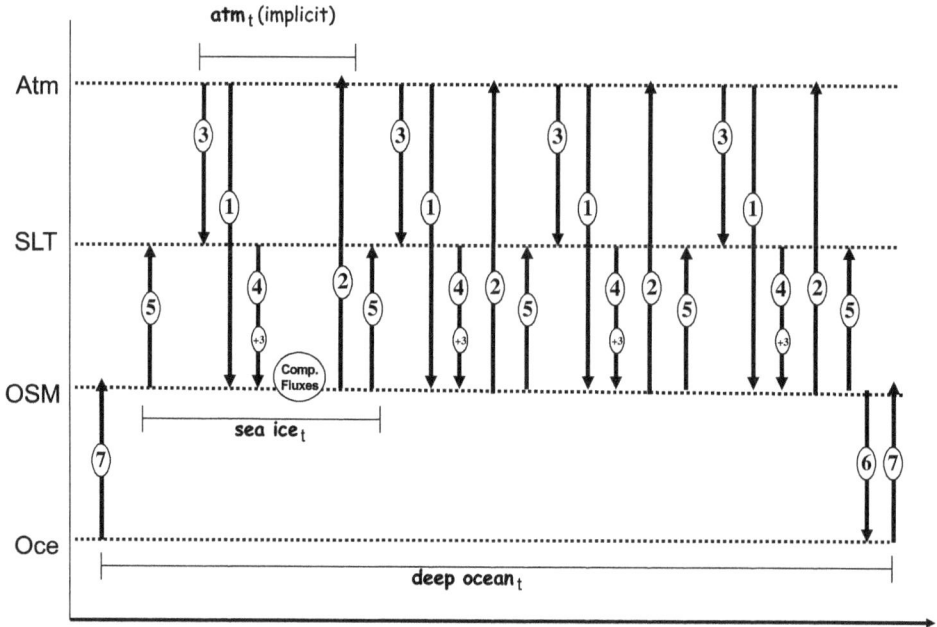

Figure 9.6. Time sequence for the coupling exchanges between the ocean (Oce), atmosphere (Atm), sea-ice (included in the Ocean surface module) via the Surface layer turbulence (SLT) and Ocean surface module OSM. (Figure from Valcke and Guilyardi [2008]. Used with permission.)

Therefore, the coupling exchanges should follow (with Fx being the frequency of exchange x): $F7 = F6 \leq F5 = F3 = F1 = F4 = F2$. The coupling exchanges (1), (2), (3), (4) and (5) can be performed many times while only one (6) and one (7) exchanges take place.

9.4 Discussion on specific coupling features and their implementation

The coupling exchanges between the component of atmosphere-ocean coupled models, including sea-ice and land components, necessarily represent a discretized implementation of the exchanges of energy, mass and momentum occurring continuously in the real Earth system. Examples presented in the previous sections show that different implementations can generate inconsistencies in time and in space.

9.4.1 *Component sequencing and time inconsistency*

Running components of a CGCMs in a concurrent mode while exchanging coupling fields at specified frequencies can lead to inconsistencies in time. For example, in the asynchronous coupling typical of European CGCMs (see

Fig. 9.1), the ocean (atmospheric) model for one coupling period uses fluxes (surface variables) that were produced by the atmospheric (ocean) model over the previous period.

The ECMWF implementation partially addresses this inconsistency by running different component models sequentially over one coupling period (see Fig. 9.2). Therefore, the ocean and the wave models use, for each period, the coupling fields calculated by the atmosphere for the same period.

In CESM2 and CMCC-CM2 (see Fig. 9.4), the land, ice and flux coupler calculate turbulent fluxes, which are aggregated and used by the atmosphere for the same coupling period. However, turbulent fluxes used in the ocean are calculated by the flux coupler over the previous period. For the ice/ocean heat and salt fluxes and penetrative shortwave fluxes, calculated by the ice, and for the shortwave and longwave downward fluxes, rain and snow, calculated by the atmosphere, the shift is even of 2 periods (as these fields are transferred via the acc2ocn module). These shifts allow the models to run concurrently and thereby decrease the wall clock time of the simulation.

In RPN CGCM, the ocean uses the state variables provided by the atmosphere over the same period to calculate the turbulent heat, humidity and momentum fluxes; in this way fluxes and atmospheric state variables are consistent over the period in the ocean. However, from the ocean to the atmosphere, there is a shift of two coupling periods; for example, during its fourth period the atmosphere receives and uses turbulent heat and humidity fluxes and wind stresses calculated by the ocean during the second coupling period.

For the vertical diffusive fluxes, one way to resolve time inconsistencies is to use an implicit formulation for their calculation as proposed in the PRISM revised physical interface (see Sec. 9.3) and implemented in FMS (see Sec. 8.2.4).

Another way to have a consistent interface is to implement Schwarz iterations while coupling the models. The Schwarz method [Marti et al., 2021, Lemarié et al., 2015] allows for correcting the time inconsistency of asynchronous coupling, leading to a coherent ocean-atmosphere interface. Lemarié et al. [2014] show that using the Schwarz coupling method in a regional coupled model for a multi-member ensemble simulation of a tropical cyclone leads to a significant reduction in the ensemble spread in terms of cyclone trajectory and intensity, thus suggesting that a source of error linked to the asynchronous coupling has been removed.

The principle of the Schwarz method is to repeat each coupling period many times, with the same initial condition for each iteration, but with evolving boundary conditions at the ocean-atmosphere interface. Instead of using the surface variables calculated by the ocean during the previous coupling period (as in the asynchronous scheme), the atmosphere uses the surface variables calculated by the ocean for that same period but during the previous iteration. And vice-versa for the ocean that uses during each iteration the fluxes calculated by the atmosphere for the same period during the previous iteration. This is repeated until convergence of the surface variables and fluxes.

Schwarz iterations have been implemented in simulations using CNRM-CM6-1D, a single-column version of CNRM-CM6-1 [Abdel-Lathif et al., 2018]. In the atmosphere, CNRM-CM6-1D is forced by horizontal advections of temperature and moisture and the large-scale vertical velocity; some variables such as the horizontal velocities, temperature and humidity are nudged toward observations or reanalysis. In the ocean, no lateral forcing or nudging is applied. The simulation performed for one point in the Indian ocean covers one day (November 13, 2011) of the Cindy Dynamo campaign [Ciesielski et al., 2014] from which the atmospheric lateral forcings are obtained. Runs with coupling periods of 300 s (which is the timestep of the models), 3,600 s, 3 hrs, 6 hrs and 12 hrs were realized. Figure 9.7 shows the resulting diurnal cycle of the Sea Surface Temperature (SST) obtained with the different coupling periods for (a) traditional asynchronous coupling and (b) after convergence of Schwarz iterations.

These results show that the coupling period can have a strong impact on asynchronous simulations (Fig. 9.7(a)). The longer the coupling period is, the more lagged the diurnal cycle is. For a coupling period of 12 hrs, the cycle is even completely inversed. In practice, a coupling period of 12 hrs is never used as this would mean that each model uses during the day coupling fields calculated by the other model during the night and vice versa. As shown in Fig. 9.7(b), the Schwarz method is very efficient to reposition the diurnal cycle.

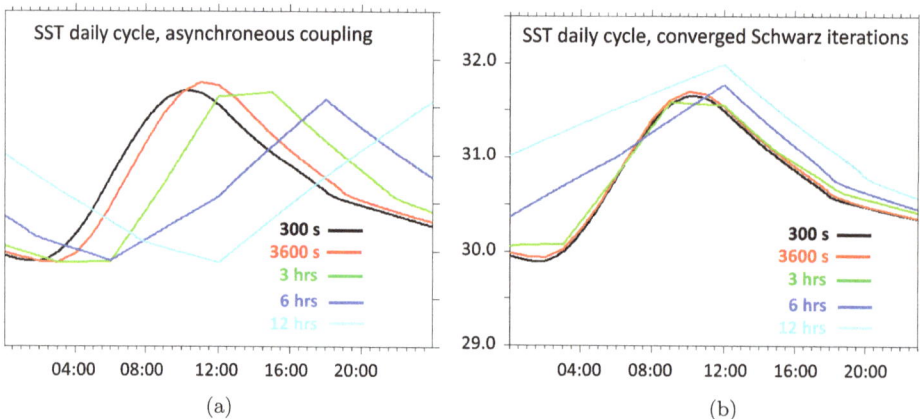

Figure 9.7.   Diurnal cycle for a one-day simulation realized with CNRM-CM-1D with different coupling periods of 300 s, 3600 s, 3 hrs, 6 hrs and 12 hrs: (a) traditional asynchronous coupling, (b) after convergence of Schwarz iterations.

It even succeeds in reversing the diurnal cycle obtained for the run with the coupling period of 12 hrs. In the current case, 20 iterations were realized for each coupling period but it was observed that convergence is always obtained in less than 10 iterations and that most of the correction is achieved after two iterations only.

The Schwarz iterative method represents an efficient way of correcting the inconsistencies introduced by the asynchronous coupling and obtaining a coherent ocean-atmosphere interface. However, the cost involved is clearly very high for 3D models as applying even only two iterations would double the cost of the simulation. Work is underway to identify subsystem in the models, e.g., only the atmospheric physics and not the whole dynamics, onto which the iterations could be applied, thereby reducing the cost of the method. Schwarz iterations can also be considered as a method to provide a clean reference coupled solution that can be used to evaluate the biases of other coupling methods.

9.4.2 *Regridding and spatial inconsistencies*

9.4.2.1 *Flux computation*

Component models assembled in a CGCM can have different grids, especially for the ocean and the atmosphere because of the different environments they represent (e.g., in an ocean model, the grid convergence singularity can be conveniently displaced over a continent). Furthermore, in current CGCMs, the ocean is usually run with a finer resolution than the atmosphere, as the synoptic scales in the ocean (e.g., ocean eddies of the $O(100)$ kms) are much smaller than the synoptic scales in the atmosphere (e.g., typical mid-latitude depressions of the $O(1000)$ kms). For example, in CNRM-CM6-1, one atmospheric cell typically contains between 5 to 10 ocean cells in the mid-latitudes, about 20 between 7°S and 7°N, and up to 40 near the South and North poles. Therefore, choices have to be made regarding the regridding of coupling fields. As turbulent surface fluxes are not linear with respect to surface variables, it is desirable that those fluxes be computed at the finest resolution of the component models involved.

This approach is applied in the RPN CGCM described in Sec. 9.2.3. In this model, surface variables such as wind and temperature are passed to the ocean model, which computes the sensible and latent fluxes and wind stress over water and ice. This calculation is therefore done at the finer resolution of the ocean. The calculation of longwave radiative flux, however, is done in the atmosphere separately over water and ice using the surface temperature regridded on the atmospheric grid. A similar approach is adopted in CESM2 and CMCC-CM2 (see Sec. 9.2.4). The PRISM physical interface presented in Sec. 9.3 also follows this approach, with the calculation of turbulent fluxes done in the flux coupler at the finest resolution.

Finally, spatial consistency is also improved in FMS with the adoption of the exchange grid, defined by the intersection of the atmospheric and surface model grids (see Sec. 8.2.4). Indeed, calculating the fluxes on the exchange grid ensures that the values of the state and surface variables used are the ones considered in the surface and atmosphere models themselves without averaging or interpolating.

However, this principle is not always applicable, especially for radiative fluxes because these fluxes depend on AGCM variables over the whole atmospheric column. So radiative fluxes are in practice always calculated in the atmosphere model, even if the ocean has a higher resolution, with ocean surface variables regridded from the ocean to the atmosphere grid.

9.4.2.2 *Sea-land mask issue*

Another source of spatial inconsistency is linked to the sea-land mask issue. Ocean and atmosphere models usually associate a specific type, ocean or land, to their grid cells; this defines the so-called sea-land mask identifying the oceanic and continental parts of the grid. Most atmospheric models are also able to consider multiple sub surfaces in each grid cell, e.g., water, ice, vegetation, urban and rock. Each sub surface is associated with a certain fraction of the total cell surface but not with a physical location in the cell.

Special care has to be taken with the conservative remapping between grids with different sea-land masks. To set up a consistent atmosphere-ocean system and have a well-posed coupled problem, the following best practice, defining coherent sea-land masks and sea fractions, should be adopted, but is applicable only if the atmosphere model can consider at least water and land sub surfaces. The original sea-land mask of the ocean model should be taken as is. For the atmosphere model, the fraction of water in each cell should be defined by the conservative remapping of the ocean mask on the atmospheric grid. A simple example is illustrated in Fig. 9.8. Thus the atmospheric coupling mask should be adapted associating a valid/active index to cells containing at least a fraction of sea. This method ensures that the total sea and land surfaces are the same in the ocean and atmosphere models, allowing global conservation of sea or land integrated quantities.

If it is not possible to define cell fractions in the atmosphere model, a mismatch necessarily occurs between the atmosphere and the ocean in their masked and non-masked total surfaces. Different normalization options can be used but the problem remains ill-posed. If the target cell area intersected by non-masked source cells is used, the area-integrated value of the coupling field will

L: masked (land) cells
S: non-masked (sea) cells

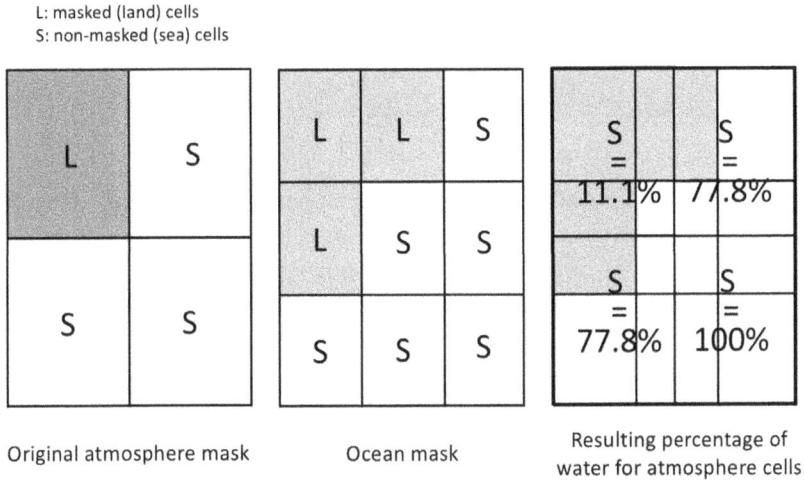

Original atmosphere mask   Ocean mask   Resulting percentage of water for atmosphere cells

Figure 9.8. Best practice for sea-land mask coherency. The ocean mask is not modified. The fraction of water in each atmospheric cell is defined by the conservative remapping of the ocean mask on the atmospheric grid; it is respectively 11.1% for the upper left cell, 77.8% for the upper right and lower left cells and 100% for the lower right cell. All atmospheric cells that contain a fraction of sea are redefined as non-masked. (Figure created by author)

not be conserved locally but field values will be realistic. If the whole target cell area is used, the local conservation of the area-integrated value of the coupling field will be ensured but unrealistic field values may be generated. Usually the former approach is preferred and a global conservation of the coupling field may be applied afterwards.

9.4.2.3 *Merging and cell fractions*

As briefly mentioned in Sec. 8.3.1.2, merging coupling fields may be required in some coupled systems. The merging may require weighting the different source fields by fractions of different sub surfaces and the problem is even further complicated if the fractions evolve with time. For instance, heat fluxes calculated over ocean cells with time-evolving sea-ice fractions need to be weighted by appropriate dynamically-varying fractions of ice and water to create a consistent merged flux in the atmosphere model. Managing fractions properly is therefore critical to conservation. As proposed in the PRISM interface (see paragraph "Subgrid scale computations" in Sec. 9.3.4), fractions should be exchanged as coupling field and synchronized in time with the fields they are going to be applied to. Inconsistencies may arise with asynchronous coupling (see Sec. 9.4.1) especially when ice fractions in one component go from nonzero to zero or vice versa during the coupling period, while the other component considers during that period fluxes calculated with fractions valid over the previous coupling period.

9.5 Perspectives

We have addressed in this chapter a few outstanding aspects related to the practical implementation of ocean-atmosphere coupling in Earth System Models. Despite years of development, the modeling of the ocean-atmosphere interface in particular remains an area of active research. This includes a wide range of aspects from the parameterizations of atmospheric and oceanic boundary layers to time-space numerical schemes going through the estimation of boundary fluxes with bulk formulae and their regridding at the air-sea interface.

In particular, ongoing research efforts are devoted to the representation of up-to-now neglected subgrid scale phenomena, such as waves, wind variability, oceanic warm layers, or SST fronts. For example, Couvelard et al. [2020] analyze the impact of ocean-wave coupling on upper-ocean properties (temperature and currents) and mixed-layer depth at global eddying scales. Blein et al. [2020] study the subgrid scale contribution to air-sea turbulent fluxes at a 20–200 km scale, showing that subgrid dynamical processes are the main contributors, through heterogeneities in the wind magnitude and direction (the so-called gustiness) and through the wind speed subgrid variance. Rackow and Juricke [2020] address the representation of the subgrid scale variability of ocean-atmosphere fluxes introducing a new flow-dependent distribution sampling (FDDS) scheme, the objective being to recreate the impact of the subgrid-scale structures, such as oceanic fronts, tropical cyclones, circulation induced by islands or coastlines.

Another important aspect that has to be considered with care, and that we have not addressed, is the numerical stability of the coupled system, which depends on the numerical schemes used in each component and on the coupling algorithm. It has been shown for instance that coupling two components with unconditional numerical stability is not a sufficient condition to guarantee the stability of the coupled system. Such issues are addressed, for example, in Lemarié et al. [2015] and Beljaars et al. [2017].

CGCMs and their expansion in scope to Earth System Models represent today key tools to explore the behavior of the Earth climate and its evolution. It is widely recognized that these models have huge capabilities and potential of useful applications for society at large. It is also recognized that these models are far from being perfect. The research community is constantly working to improve them and we can anticipate that such efforts will continue for years to come.

Bibliography

Adcroft, A. J. and Hallberg, R. W. (2006) On methods for solving the oceanic equations of motion in generalized vertical coordinates. *Ocean Modelling*, 11, 224–233.

Alexander, M. and Deser, C. (1995) A mechanism for the recurrence of wintertime midlatitude SST anomalies. *J. Phys. Oceanogr.*, 25, 122–137.

Alexander, M., Deser, C. and Timlim, M. (1999) The reemergence of SST anomalies in the North Pacific Ocean. *J. Climate*, 12, 2419–2433.

An S-I, Hsieh, W. W. and Jin, F.-F. (2005) A nonlinear analysis of the ENSO cycle and its interdecadal changes. *J. Climate*, 18, 3229–3239.

An, S.-I. and Kang, I.-S. (2000) A further investigation of the recharge oscillator paradigm for ENSO using a simple coupled model with the zonal mean and eddy separated, *J. Climate*, 13, 1987–1993.

An, S.-I. (2009) A review of interdecadal changes in the nonlinearity of the El Niño-Southern Oscillation. *Theor. App. Climatology*, 97(1–2), 29–40. https://doi.org/10.1007/s00704-008-0071-z.

An, S.-I. (2008) Interannual variations of the tropical ocean instability wave and ENSO. *J. Climate*, 21, 3680–3686

An, S.-I. and Jin, F.-F. (2004) Nonlinearity and asymmetry of ENSO. *J. Climate*, 17, 2399–2412.

An, S.-I. and Kim, J.-W. (2017) Role of nonlinear ocean dynamic response to wind on the asymmetrical transition of El Niño–La Niña. *Geophys. Res. Lett.*, 44, 393–400.

An, S.-I. and Wang, B. (2000) Interdecadal change of the structure of the ENSO mode and its impact on ENSO frequency. *J. Climate*, 13, 2044–2055.

An, S.-I., Kim, J.-W., Im, S.-H., Kim, B.-M., and Park, J.-H. (2012) Recent and future sea surface temperature trends in tropical Pacific warm pool and cold tongue regions, *Climate Dynamics*, 39, 1373–1383.

An, S.-I., Tziperman, E., Okumura, Y., and Li, T. (2020) Irregularity and asymmetry. In *El Niño Southern Oscillation in a Changing Climate* (eds. M. McPhaden, A. Santoso and W. Cai). American Geophysical Union, 528pp, ISBN: 978-1-119-54815-7.

An, S.-I., Wang, C. and Mechoso, C. R. (2021) Teleconnections in the Atmosphere. In *Interacting Climates of Ocean Basins* (ed. C. R. Mechoso). Cambridge University Press, Cambridge, UK and New York, NY, USA. 358 pp, ISBN: 9781108492706.

An, S.-I. and Im, S.-H. (2014) Blunt ocean dynamical thermostat in response of tropical eastern Pacific SST to global warming. *Theoretical and Applied Climatology*, 118, 173–183.

An, S.-I. and Jin, F.-F. (2001) Collective role of thermocline and zonal advective feedbacks in the ENSO mode. *J. Climate*, 14, 3421–3432.

An, S.-I. and Kim, J.-W. (2018) ENSO transition asymmetry: Internal and external causes and intermodel diversity. *Geophys. Res. Lett.*, 45, 5095–5104.

Arakawa, A. and Konor, C. S. (1996) Vertical differencing of the primitive equations based on the Charney–Phillips grid in hybrid sigma–p vertical coordinates. *Mon. Wea. Rev.*, 124, 511–528.

Arakawa, A. and Lamb, V. R. (1977) Computational design of the basic dynamical processes of the UCLA general circulation model. *Methods in Computational Physics: Advances in Research and Applications*, 17, 173–265. doi: 10.1016/B978-0-12-460817-7.50009-4.

Ashok, K., Behera, S. K., Rao, S. A., Weng, H., and Yamagata, T. (2007) El Niño Modoki and its possible teleconnection. *J. Geophys. Res. Oceans*, 112, C11007.

Avissar, R. (1992) Conceptual aspects of a statistical-dynamical approach to represent landscape subgrid-scale heterogeneities in atmospheric models, *J. Geophys. Res.*, 97, 2729–2742.

Avissar, R. and Pielke, R. (1989) A parameterization of heterogeneous land surfaces for atmospheric numerical models and its impact on regional meteorology, *Mon. Wea. Rev.*, 117, 2113–2136.

Balaji, V., Anderson, J., Held, I., Winton, M., Durachta, J., Malyshev, S., and Stouffer, R. J. (2006) The exchange grid: A mechanism for data exchange between Earth System components on independent grids. In *Proceedings of the 2005 International Conference on Parallel Computational Fluid Dynamics*, College Park, MD, USA, pp. 179–186, https://doi.org/10.1016/B978-044452206-1/50021-5.

Balmaseda, M. A., Anderson, D. L. T. and Davey, M. K. (1994) ENSO prediction using a dynamical ocean model coupled to statistical atmospheres. *Tellus*, 46(A) 4, 497–511

Balmaseda, M. A., Mogensen, K. and Weaver, A. T. (2013) Evaluation of the ECMWF ocean reanalysis system ORAS4. *Q. J. R. Meteorol. Soc.*, 139, 1132–1161. doi: 10.1002/qj.2063

Barnett, T. P., Latif, M., Graham, N., Flugel, M., Pazan, S., and White, W. (1993) ENSO and ENSO-related predictability. Part I: Prediction of equatorial Pacific sea surface temperature with a hybrid coupled ocean–atmosphere model. *J. Climate*, 6, 1545–1566.

Barnston, A. G., Tippett, M. K., L'Heureux, M. L., Li, S., and Dewitt, D. G. (2012) Skill of real-time seasonal ENSO model predictions during 2002–11: Is our capability increasing? *Bull. Amer. Meteor. Soc.*, 93, 631–51.

Battisti, D. S. (1988) The dynamics and thermodynamics of a warming event in a coupled tropical atmosphere-ocean model. *J. Atmos. Sci.*, 45, 2889–2919.

Battisti, D. S. and A. C. Hirst (1989) Interannual variability in the tropical atmosphere/ocean system: Influence of the basic state, ocean geometry and nonlinearity. *J. Atmos. Sci.*, 46, 1687–1712.

Behringer, D. W. and Xue, Y. (2004) Evaluation of the global ocean data assimilation system at NCEP: The Pacific Ocean. Eighth Symposium on Integrated Observing and Assimilation Systems for Atmosphere, Oceans, and Land Surface, AMS 84th Annual Meeting, Washington State Convention and Trade Center, Seattle, Washington, pp. 11–15.

Behringer, D.W., Ji, M. and Leetmaa, A. (1998) An improved coupled model for ENSO prediction and implications for ocean initialization. Part I: The ocean data assimilation system. *Mon. Wea. Rev.*, 126, 1013–1021.

Beljaars, A., Dutra, E., Balsamo, G., and Lemarie, F. (2017) On the numerical stability of surface-atmosphere coupling in weather and climate models. *Geosci. Model Dev.*, 10, pp. 977–989, https://doi.org/10.5194/gmd-10-977-2017.

Bellucci, A., Gualdi, S. and Navarra, A. (2010) The double-ITCZ syndrome in coupled general circulation models: The role of large-scale vertical circulation regimes. *J. Climate*, 23, 1127–145, https://doi.org/10.1175/2009JCLI3002.1.

Bitz, C. M., Shell, K. M., Gent, P. R., Bailey, D. A., Danabasoglu, G., Armour, K. C., Holland, M. M., and Kiehl, J. T. (2012) Climate sensitivity of the Community Climate System Model, Version 4. *J. Climate*, 25, 3053–3070, https://doi.org/10.1175/JCLI-D-11-00290.1.

Bjerknes, J. (1966) A possible response of the atmospheric Hadley circulation to equatorial anomalies of ocean temperature. *Tellus*, 18, 820–829.

Bjerknes, J. (1969) Atmospheric teleconnection from the equatorial Pacific. *Mon. Wea. Rev.*, 97, 163–172. https://doi.org/10.1175/1520-0493.

Blackadar, A. K. and Tennekes, H. (1968) Asymptotic similarity in neutral barotropic planetary boundary layers. *J. Atmos. Sci.*, 25, 1015–1020.

Bleck, R. (2002) An oceanic general circulation model framed in hybrid isopycnic-Cartesian coordinates. *Ocean Modelling*, 4, 55–88.

Blein, S., Roehrig, R., Voldoire, A., and Faure, G. (2020) Meso-scale contribution to air-sea turbulent fluxes at GCM scale, *Q. J. Roy. Meteor. Soc.*, 146(730), 2466–2495. https://doi.org/10.1002/qj.3804.

Bogenshutz, P. A., Gettelman, A., Morrison, H., Larson, V. E., Schanen, D. P., Meyer, N. R., and Craig, C. (2012) Unified parameterization of the planetary boundary layer and shallow convection with a higher-order turbulence closure in the Community Atmosphere Model: Single column experiments. *Geosci. Model Dev.*, 5, 1407–1423.

Bogenshutz, P. A., Gettelman, A., Morrison, H., Larson, V. E., Schanen, D. P., Meyer, N. R., and Craig, C. (2013) Higher-order turbulence closure and its impact on climate simulations in the Community Atmosphere Model. *J. Climate*, 26(23), 9655–9676.

Bretherton, C. S. and Park, S. (2009) A new moist turbulence parameterization in the community atmosphere model. *J. Climate*, 22, 3422–3448.

Bretherton, C. S., Smith, C. and Wallace, J. M. (1992) An intercomparison of methods for finding coupled patterns in climate data. *J. Climate*, 5, 541–560.

Brown, A., Milton, S., Cullen, M., Golding, B., Mitchell, J., and Shelly, A. (2012) Unified modeling and prediction of weather and climate: A 25-year journey. *Bull. Amer. Meteor. Soc.*, 93 (12), 1865–1877. https://doi.org/10.1175/BAMS-D-12-00018.1.

Bryan, K. (1969) A numerical method for the study of the circulation of the world ocean. *J. Comp. Phys.*, 4, 347–376.

Budyko, M. I. (1969) The effect of solar radiation variations on the climate of the earth. *Tellus*, 21, 611–619.

Burgers, G. (1999) The El Niño stochastic oscillator. *Climate Dyn.*, 15, pp. 521–531.

Burgers, G. and Stephenson, D. B. (1999) The "normality" of El Niño. *Geophys. Res. Lett.*, 26, 1027–1030.

Businger, J. A., Wyngaard, J. C., Izumi, Y., and Bradley, E. F. (1971) Flux-profile relationships in the atmospheric surface layer. *J. Atmos. Sci.*, 28, 181–189.

Cai, W., Borlace, S., Lengaigne, M., van Rensch, P., Collins, M., Vecchi, G., Timmermann, A., Santoso, McPhaden, M. J., Wu, L., England, M. H., Wang, G., Guilyardi, E., and Jin. F.-F. (2014) Increasing frequency of extreme El Niño events due to greenhouse warming. *Nature Climate Change*, 4, 111–116.

Cai, W., Wang, G., Dewitte, B., Wu, L., Santoso, A., Takahashi, K., Yang, Y., Carréric, A., and McPhaden, M. J. (2018) Increased variability of eastern Pacific El Niño under greenhouse warming. *Nature*, 564, 201–206.

Cane, M. A., Zebiak, S. E. and Dolan, S. C. (1986) Experimental forecasts of El Niño. *Nature*, 321, 827–832.

Caron, L.-P., Hermanson, L., Dobbin, A., Imbers, J., Lledó, L., and Vecchi, G. A. (2017) How skillful are the multiannual forecasts of Atlantic hurricane activity? *Bull. Amer. Meteor. Soc.*, 99(2), 403–413.

Castaño-Tierno, A., Mohino, E., Rodríguez-Fonseca, B., and Losada, T. (2018) Revisting the CMIP5 thermocline in the equatorial Pacific and Atlantic Oceans. *Geophy. Res. Lett.*, 45, 12,963–12,971.

Cazenave, A., Meehl, G., Montoya, M., Toggweiler, J. R., and Wieners, C. (2021) Climate change and impacts on variability and interactions. In *Interacting Climates of Ocean Basins* (ed. C. R. Mechoso). Cambridge University Press, Cambridge, UK and New York, NY, USA.

Chang, P. and Philander, S. G. (1994) A coupled ocean-atmosphere instability of relevance to the seasonal cycle. *J. Atmos. Sci.*, 51, 3628–3648.

Chang, P., Ji, L. and Saravanan, R. (2001) A hybrid coupled model study of tropical Atlantic variability. *J. Climate*, 14, 361–390

Chang, P., Ji, L. and Li, H. (1997) A decadal climate variation in the tropical Atlantic Ocean from thermodynamic air-sea interactions. *Nature*, 385, 516–518.

Charney, J. G., Arakawa, A., Baker, D. J., Bolin, B., and Dickinson, R. E. (1979) Carbon dioxide and climate: A scientific assessment. National Academy of Sciences Tech. Rep., 34 pp., doi:10.17226/12181.

Charnock, H. (1955) Wind stress on a water surface. *Quart. J. Roy. Meteor. Soc.*, 81, 639–640.

Chase, T. N., Pielke, R. A. Sr., Kittel, T. G. F., Nemani, R., and Running, S. W. (1996) Sensitivity of a general circulation model to global changes in leaf area index. *J. Geophys. Res.*, 101, 7393–7408.

Chen, C., Cane, M. A., Wittenberg, A. T., and Chen, D. (2017) ENSO in the CMIP5 simulations: Life cycles, diversity, and responses to climate change. *J. Climate*, 30, 775–801.

Chen, D. and Cane, M. A. (2008) El Niño prediction and predictability, *J. Comput. Phys.*, 227, 3625–3640.

Chen, D., Cane, M. A., Kaplan, A., Zebiak, S. E., and Huang, D. (2004) Predictability of El Niño over the past 148 years. *Nature*, 428, 733–736

Chen, D., Cane, M. A., Zebiak, S. E., and Kaplan, A. (1998) The impact of sea level data assimilation on the Lamont model prediction of the 1997/98 El Niño. *Geophys. Res. Lett.*, 25, 2837–2840.

Chen, D., Zebiak, S. E., Busalacchi, A. J., and Cane, M. A. (1995) An improved procedure for El Niño forecasting: Implications for predictability. *Science*, 269, 1699–1702, doi:10.1126/science.269.5231.1699.

Chen, H.-C., Tseng, Y. H., Hu, Z.-Z., and Ding, R. (2020) Enhancing the ENSO predictability beyond the spring barrier. *Sci. Reports*, 10:984, doi.org/10.1038/s41598-020-57853-7.

Chen, M., Li, T., Shen, X., and Wu, B. (2016) Relative roles of dynamic and thermodynamic processes in causing evolution asymmetry between El Niño and La Niña. *J. Climate*, 29(6), 2201–2220.

Cheng, W., Chiang, J. C. and Zhang, D. (2013) Atlantic meridional overturning circulation (AMOC) in CMIP5 models: RCP and historical simulations. *J. Climate*, 26, 7187–7197, https://doi.org/10.1175/JCLI-D-12-00496.1.

Chi, J., Du, Y., Zhang, Y., Nie, X., Shu, P., and Qu, T. (2019) A new perspective of the 2014/15 failed El Niño as seen from ocean salinity. *Science Rep.*, 9, 2730, https://doi.org/10.1038/s41598-019-38743-z.

Christensen, J. H., Kumar, K. K., Aldrian, E., An, S.-I., Cavalcanti, I. F. A., de Castro, M. Dong, W. Goswami, P., Hall, A., Kanyanga, J. K., Kitoh, A., Kossin, J., Lau, N.-C., Renwick, J., Stephenson, D. B., Xie, S.-P., and Zhou, T. (2013) Climate phenomena and their relevance for future regional climate change. *Climate Change 2013: The Physical Science Basis*, T. F. Stocker *et al.* (eds.), Cambridge University Press, pp. 1217–1308.

Choi, K. Y., Vecchi, G. A. and Wittenberg, A. T. (2013) ENSO transition, duration, and amplitude asymmetries: Role of the nonlinear wind stress coupling in a conceptual model. *J. Climate*, 26(23), 9462–9476.

Ciesielski, P. E., Hungjui Yu, H., Johnson, R. H., Yoneyama, K., Katsumata, M., Long, C. N., Wang, J., Loehrer, S. M., Young, K., Williams, S. F., Brown, W., Braun, J., and Van Hove, T. (2014). Quality-controlled upper-air sounding dataset for DYNAMO/CINDY/AMIE: Development and corrections. *J. Atmos. Ocean. Tech.*, 31(4), 741–764.

Cimatoribus, A., Drijfhout, S. S. and Dijkstra, H. A. (2012a) A global hybrid coupled model based on atmosphere-SST feedbacks. *Climate Dyn.*, 38, 745–760.

Cimatoribus, A., Drijfhout, S. S., den Toom, M., and Dijkstra, H. A. (2012b) Sensitivity of the Atlantic meridional overturning circulation to South Atlantic freshwater anomalies. *Climate Dyn.*, 39, 2291–2306

Clement, A. C., Seager, R., Cane, M. A., and Zebiak, S. E. (1996) An ocean dynamical thermostat, *J. Climate*, 9, 2190–2196.

Collins, M., An, S.-I., Cai, W., Ganachaud, A., Guilyardi, E., Jin, F.-F., Jochum, M., Lengaigne, M., Power, S., Timmermann, A., Vecchi, G., and Wittenberg, A. (2010) The impact of global warming on the tropical Pacific ocean and El Niño. *Nature Geosciences*, 3, 391–397.

Collins, N., Theurich, G., DeLuca, C., Suarez, M., Trayanov, A., Balaji, V., Li, P., Yang, W., Hill, C., and da Silva, A. (2005) Design and Implementation of Components in the Earth System Modeling Framework. *Int. J. High Perfor. Comput. Apps.*, 19(3), 341–350.

Corti, S., Palmer, T., Balmaseda, M., Weisheimer, A., Drijfhout, S., Dunstone, N., Christensen, J. H., Krishna Kumar, K., Aldrian, E., An, S.-I., Cavalcanti, I. F. A., de Castro, M., Dong, W., Goswami, P., Hall, A., Kanyanga, J. K., Kitoh, A., Kossin, J., Lau, N.-C., Renwick, J., Stephenson, D. B., Xie, S.-P., and Zhou, T. (2013) Climate phenomena and their relevance for future regional climate change. In: *Climate Change 2013: The Physical Science Basis*. Contribution of Working Group I to the Fifth Assessment Report of the Intergovernmental Panel on

Climate Change [Stocker, T.F., D. Qin, G.-K. Plattner, M. Tignor, S.K. Allen, J. Boschung, A. Nauels, Y. Xia, V. Bex and P.M. Midgley (eds.)]. Cambridge University Press.

Couvelard, X., Lemarié, F., Samson, G., Redelsperger, J.-L., Ardhuin, F., Benshila, R., and Madec, G. (2020) Development of a two-way-coupled oceanwave model: assessment on a global NEMO(v3.6)WW3(v6.02) coupled configuration. *Geosci. Model Dev.*, 13, pp. 3067–3090, https://doi.org/10.5194/gmd-13-3067-2020

Cox, M. D. (1984) A primitive equation three-dimensional model of the ocean. GFDL Ocean Group Tech. Rept. No. 1, GFDL/NOAA, Princeton University, Princeton, NJ, 250 pp.

Craig, A. P., Jacob R., Kauffman B., Bettge T., Larson J., Ong E., Ding, C., and He, Y. et al. (2005) CPL6: The new extensible high performance parallel coupler for the Community Climate System Model. *I. J. High Perfor. Comput. Apps.*, 19, 309–328.

Craig, A. P., Vertenstein, M. and Jacob, R. (2012) A New Flexible Coupler for Earth System Modeling developed for CCSM4 and CESM1, *Int. J. High Perform. Comput. Apps.*, 26(1), 31–42, https://doi.org/10.1177/1094342011428141.

Craig, A., Valcke, S. and Coquart, L. (2017) Development and performance of a new version of the OASIS coupler, OASIS3-MCT_3.0, *Geosci. Model Dev.*, 10, 3297–3308, https://doi.org/10.5194/gmd-10-3297-2017.

Da Silva, A., Young, A. C. and Levitus, S. (1994) *Anomalies of Fluxes of Heat and Momentum.* Vol. 3, *Atlas of Surface Marine Data 1994*, NOAA Atlas NESDIS 8, 411 pp.

Danabasoglu, G. and Gent, P. R. (2009) Equilibrium climate sensitivity: Is it accurate to use a slab ocean model? *J. Climate*, 22, 2494–2499.

Danabasoglu, G., Lamarque, J.-F., Bacmeister, J., Bailey, D. A., DuVivier, A. K., Edwards, J., Emmons, L. K., Fasullo, J., Garcia, R., Gettelman, A., Hannay, C., Holland, M. M., Large, W. G., Lauritzen, P. H., Lawrence, D. M., Lenaerts, J. T. M., K. Lindsay, K., Lipscomb, W. H., Mills, M. J., Neale, R., Oleson, K. W., Otto–Bliesner, B., Phillips, A. S., Sacks, W., Tilmes, S., van Kampenhout, L., Vertenstein, M., Bertini, A., Dennis, J., Deser, C., Fischer, C., Fox–Kemper, B., Kay, J. E., Kinnison, D., Kushner, P. J., Larson, V. E., Long, M. C., Mickelson, S., Moore, J. K., Nienhouse, E., Polvani, L., Rasch, P. J., and Strand W. G. (2020) The Community Earth System Model Version 2 (CESM2). *J. Adv. in Modeling Earth Systems*, 12, e2019MS001916. https://doi.org/10.1029/2019MS001916.

Danilov, S. (2013) Ocean modeling on unstructured meshes. *Ocean Mod.*, 69, 195–210. https://doi.org/10.1016/j.ocemod.2013.05.005.

Davey, M. K., Huddleston, M., Sperber, K. R., Braconnot, P., Bryan, F., Chen, D., Colman, R. A., Cooper, C., Cubasch, U., Delecluse, P., DeWitt, D., Fairhead, L., Flato, G., Gordon, C., Hogan, T., Ji, M., Kimoto, M., Kitoh, A., Knutson, T. R., Latif, M., Le Treut, H., Li, T., Manabe, S., Mechoso, C. R., Meehl, G. A., Power, S. B., Roeckner, E., Terray, L., Vintzileos, A., Voss, R., Wang, B., Washington, W. M., Yoshikawa, I., Yu, J.-Y., Yukimoto, S., and Zebiak, S. E. (2002) STOIC:Astudy of coupled model climatology and variability in tropical ocean regions. *Climate Dyn.*, 18, 403–420, https://doi.org/10.1007/s00382-001-0188-6.

DeMott, C. A., Klingaman, N. P. and Woolnough, S. J. (2015) Atmosphere-ocean coupled processes in the Madden-Julian Oscillation. *Rev. Geophys.*, 53, 1099–1154.

Deardorff, J. W. (1972) Theoretical expression for the countergradient vertical heat flux. *J. Geophys. Res., Oceans and Atmospheres*, 77(30), 5900–5904.

Dee, D. P., Uppala, S. M., Simmons, A. J., Berrisford, P., Poli, P., Kobayashi, S., Andrae, U., Balmaseda, M. A., Balsamo, G., Bauer, P., Bechtold, P., Beljaars, A. C. M., van de Berg, L. Bidlot, J., Bormann, N., Delsol, C., Dragani. R., Fuentes, M., Geer, A. J., Haimberger, L., Healy, S. B., Hersbach, H., Hólm, E. V., Isaksen, L., Kållberg, P., Köhler, M., Matricardi, M., McNally, A. P., Monge–Sanz, B. M., Morcrette, J.-J., Park, B.-K., Peubey, C., de Rosnay, P., Tavolato, C., Thépaut, J.-N., and Vitart, F. (2011) The ERA-Interim reanalysis: configuration and performance of the data assimilation system. *Q. J. R. Meteor. Soc.*, 137, 553–597, DOI: 10.1002/qj.828.

Deppenmeier, A. L., Haarsma, R. J. and Hazeleger, W. (2016): The Bjerknes feedback in the tropical Atlantic in CMIP5 models. *Climate Dyn.*, 47, 2691–2707.

Deser, C. and Wallace, J. M. (1987) El Niño events and their relation to the southern oscillation: 1925–1986. *J. Geophys. Res. Ocean*, 92, pp. 14189–14196 (1987).

Dewitte, B. and Perigaud, C. (1996) El Niño–La Niña events simulated with Cane and Zebiak's model and observed with satellite or in situ data. Part II: Model forced with observation. *J. Climate*, 9, 1188–1207.

Dewitte, B., Illig, S., Parent, L., du Penhoat, Y., Gourdeau, L., and Verron, J. (2003) Tropical Pacific baroclinic mode contribution and associated long waves for the 1994–1999 period from an assimilation experiment with altimetric data. *J. Geophys. Res.*, 108, 3121, doi:10.1029/2002JC001362.

Dickinson, R. E., Shaikh, M., Bryant, R., and Graumlich, L. (1998) Interactive canopies for a climate model, *J. Climate*, 11, 2823–2836.

Dirmeyer, P. A. and Shukla, J. (1994) Albedo as a modulator of climate response to tropical deforestation. *J. Geophysical Research — Atmospheres*, 99(D10), 20863–20877.

Dommenget, D. (2010) The slab ocean El Niño. *Geophys. Res. Lett.*, 37, L20701, doi:10.1029/2010GL044888.

Dommenget, D., Bayr, T. and Frauen, C. (2013) Analysis of the non-linearity in the pattern and time evolution of El Niño Southern Oscillation. *Climate Dyn.*, 40(11–12), 2825–2847. https://doi.org/10.1007/s00382-012-1475-0.

Douville, H., Chauvin F. and Broqua, H. (2001) Influence of soil moisture on the Asian and African monsoons. Part I: Mean monsoon and daily precipitation. *J. Climate*, 14, 2381–2403.

Drennan, W. M., Graber, H. C., Hauser, D., and Quentin, C. C. (2003) On the wave age dependence of wind stress over pure wind seas. *J. Geophys. Res.*, 108, 8062, doi:10.1029/2000JC000715.

Drummond, L. A., Demmel, J., Mechoso, C.R., Robinson, H., Sklower, K., and Spahr, J.A. (2001) A data broker for distributed computing environments. *Lecture Notes in Computer Science*, Vol. 2073, pp. 31–40.

Ducoudre, N. I., Laval, K. and Perrier, A. (1993) SECHIBA, a new set of parameterizations of the hydrologic exchanges at the land/atmosphere interface within the LMD atmospheric general circulation model. *J. Climate*, 6, 248–273.

Eisenman, I., Yu, L. and Tziperman, E. (2005) Westerly wind bursts: ENSO's tail rather than the dog? *J. Climate*, 18(24), 5224–5238. https://doi.org/10.1175/JC LI3588.1.

England, M. H., McGregor, S., Spence, P., Meehl, G. A., Timmermann, A., Cai, W., Gupta, A. S., McPhaden, M. J., Purich, A., and Santoso, A. (2014) Recent intensification of wind-driven circulation in the Pacific and the ongoing warming hiatus. *Nature Climate Change*, 4, 222–227.

Entekhabi, D. and Eagleson, P. S. (1989) Land surface hydrology parameterization for atmospheric general circulation models including subgrid scale spatial variability, *J. Climate*, 2(8), 816–831.

Eyring, V., Bony, S., Meehl, G. A., Senior, C. A., Stevens, B., Stouffer, R. J., and Taylor, K. E. (2016) Overview of the Coupled Model Intercomparison Project Phase 6 (CMIP6) experimental design and organization, *Geosci. Model Dev.*, 9, 1937–1958, https://doi.org/10.5194/gmd-9-1937-2016.

Famiglietti, J. and Wood, E. (1991) Evapotranspiration and runoff from large land areas: Land surface hydrology for atmospheric general circulation models. *Surv. Geophys.*, 12, 179–204.

Fedorov, A., Barreiro, M., Bocaletti, G., Pacanowski, R., and Philander, S. G. (2007) The freshening of surface waters in high latitudes: Effects on the thermohaline and wind-driven circulations, *J. Phys. Oceanogr.*, 37, 896–907.

Fox-Kemper, B., Adcroft, A., Böning, C. W., Chassignet, E. P., Curchitser, E., Danabasoglu, G., Eden, C., England, M. H., Gerdes, R., Greatbatch, R. J., Griffies, S. M., Hallberg, R. W., Emmanuel, H., Heimbach, P., Hewitt, H. T., Hill, C. N., Yoshiki, K., Legg, S., Le Sommer, J., Masina, S., Marsland, S. J., Penny, S. G., Qiao, F., Ringler, T. D., Treguier, A. M., Hiroyuki, T., Uotila, P., and Yeager, S. G. (2019) Challenges and prospects in Ocean Circulation Models. *Frontiers in Marine Science*, 6, 65pp.

Frauen, C. and Dommenget, D. (2012) Influences of the tropical Indian and Atlantic Oceans on the predictability of ENSO. *Geophys. Res. Lett.*, 39(2), L02706.

Geiss, A., Marchand, R. and Thompson, L. A. (2020) The influence of sea surface temperature reemergence on marine stratiform clouds. *Geophys. Res. Lett.*, 47, e2020GL086957.

Gelaro, R., et al. (2017) The modern-era retrospective analysis for research and applications, version 2 (MERRA-2). *J. Climate*, 30, 5419–5454.

Gent, P. and Cane, M. A. (1989) A reduced gravity, primitive equation model of the upper equatorial ocean. *J. Comput. Phys.*, 81, pp. 444–480.

Gill, A. E. (1980) Some simple solutions for heat induced tropical circulations. *Q. J. Royal Meteor. Soc.*, 106, 447–462.

Golaz, J.-C., Caldwell, P. M., Van Roekel, L. P., Petersen, M. R., Tang, Q., Wolfe, J. D., and co-authors (2019) The DOE E3SM coupled model version 1: Overview and evaluation at standard resolution. *J. Adv. Model. Earth Syst.*, 11(7), 2089–2129, https://doi.org/10.1029/2018MS001603.

Goosse, H. and Fichefet, T. (1999) Importance of ice-ocean interactions for the global ocean circulation: a model study. *J. Geophys. Res.*, 104, 23337–23355.

Goubanova, K., Sanchez-Gomez, E., Frauen, C., and Voldoire, A. (2019) Respective roles of remote and local wind stress forcings in the development of warm SST errors in the south-eastern tropical Atlantic in a coupled high-resolution model. *Climate Dyn.*, 1–24, doi:10.1007/s00382-018-4197-0.

Graham, N. E. and Barnett, T. P. (1987) Sea surface temperature, surface wind divergence, and convection over tropical oceans. *Science*, 238, pp. 657–659.

Gregory, J. M., Ingram, W. J., Palmer, M. A., Jones, G. S., Stott, P. A., Thorpe, R. B., Lowe, J. A., Johns, T. C., and Williams, K. D. (2004) A new method for diagnosing radiative forcing and climate sensitivity. *Geophys. Res. Lett.*, 31, L03205, doi:10.1029/2003GL018747.

Griffies, S. M. and Treguier, A. M. (2013) Ocean circulation models and modeling. In *Ocean Circulation and Climate*, 2nd Edition, edited by G. Siedler, S.M. Griffies, J. Gould, and J. Church.

Griffies, S. M., Adcroft, A. J., Banks, H., Boning, C. W., Chassignet, E. P., Danabasoglu, G., Danilov, S., Deleersnijder, E., Drange, H., England, M., Fox-Kemper, B., Gerdes, R., Gnanadesikan, A., Greatbatch, R. J., Hallberg, R. W., Hanert, E., Harrison, M. J., Legg, S., Little, C. M., Madec, G., Marsland, S. J., Nikurashin, M., Pirani, A., Simmons, H. L., Schroter, J., Samuels, B. L., Treguier, A.-M., Toggweiler, J. R., Tsujino, H., Vallis, G. K., and White, L. (2010) Problems and prospects in large-scale ocean circulation models, In *Proceedings of the OceanObs'09 Conference: Sustained Ocean Observations and Information for Society,* Vol. 2, eds. J. Hall, D. E. Harrison and D. Stammer (Venice: ESA Publication WPP-306).

Griffies, S. M., Adcroft, A. J. and Hallberg, R. W. (2020) A primer on the vertical Lagrangian-remap method in ocean models based on finite volume generalised vertical coordinates. *J. Adv. Mod. Earth Systems*, 12, e2019MS001954, doi:10.1029/2019MS001954.

Griffies, S. M., Böning, C., Bryan, F. O., Chassignet, E. P., Gerdes, R., Hasumi, H., Hirst, A., Treguier, A.-M., and Webb, D. (2000) Developments in ocean climate modelling. *Ocean Mod.*, 2, 123–192.

Hahn, D. G. and Manabe, S. (1975) The role of mountains in the South Asian monsoon circulation. *J. Atmos. Sci.*, 32, 1515–1541.

Hanke, M., Redler, R., Holfeld, T., and Yastremsky, M. (2016) YAC 1.2.0: new aspects for coupling software in Earth system modelling, *Geosci. Model Dev.*, 9, 2755–2769, https://doi.org/10.5194/gmd-9-2755-2016.

Harrison, D. E. and Vecchi, G. A. (1997) Westerly wind events in the tropical Pacific. *J. Climate*, 10, 3131–3156.

Harvey, L. D. D. (1986) Computational efficiency and accuracy of methods for asynchronously coupling atmosphere-ocean climate models. Part I: Testing with a seasonal cycle. *J. Phys. Oceanogr.*, 16, 11–24.

Hayashi, M. and Watanabe, M. (2017) ENSO complexity induced by state dependence of westerly wind events. *J. Climate*, 30(9), 3401–3420. https://doi.org/10.1175/J CLI-D-16-0406.1.

Hazeleger, W., Severijns, C., Semmler, T., Ştefănescu, S., Yang, S., Xueli Wang, Wyser, K., Dutra, E., Baldasano, J. M., Bintanja, R., Philippe Bougeault, P., Caballero, R., Ekman, A. M. L., Christensen J. H., van den Hurk, B., Jimenez, P., Jones, C., Kållberg, P., Koenigk, T., McGrath, R., Miranda, P., van Noije, T., Palmer, T., Parodi, J. A., Schmith, T., Selten, F., Storelvmo, T., Sterl, A., Tapamo, H., Vancoppenolle, M., Viterbo, P., and Willén, U. (2010) EC-Earth: A seamless earth system prediction approach in action. *Bull. Amer. Meteor. Soc.*, 91, 1357–1363, doi:10.1175/2010BAMS2877.1.

Hazeleger, W., Kröger, J., Pohlmann, H., Smith, D., Storch, J.-S. v., and Wouters, B. (2015) Impact of initial conditions versus external forcing in decadal climate predictions: A sensitivity experiment. *J. Climate*, 28(11), 4454–4470.

Held, I. M. and Soden, B. J. (2000) Water vapor feedback and global warming. *Annual Review of Energy and the Environment*, 25, 441–475.

Held, I. M. and Soden, R. J. (2006) Robust responses of the hydrological cycle to global warming. *J. Climate*, 19, 5686–5699.

Hersbach, H. (2010) Sea-surface roughness and drag coefficient as function of neutral wind speed. ECMWF Technical Memoranda 630, doi.10.21957/hcgkicamg.

Hill, C., DeLuca, C., Balaji, V., Suarez, M., and da Silva, A. (2004) Architecture of the Earth System Modeling Framework, *Comput. Sci. Eng.*, 6, 18–28.

Hogan, R. J. and Bozzo, A. (2016) ECRAD: A new radiation scheme for the IFS (Tech. Memo. No. 787): ECMWF. 33 pp.

Holton, J. R. (2012) *An Introduction to Dynamic Meteorology.* 5th ed. Academic Press., 552 pp.

Holtslag, A. A. M., De Bruijn, E. I. F. and Pan, H.-L. (1990) A high resolution air mass transformation model for short-range weather forecasting. *Mon. Wea. Rev.*, 118, 1561–1575.

Holtslag, B. and Boville, B. A. (1993) Local versus nonlocal boundary-layer diffusion in a global model. *J. Climate*, 6, 1825–1842

Hu, S. and Fedorov, A. V. (2019) The extreme El Niño of 2015–2016: The role of westerly and easterly wind bursts and preconditioning by the failed 2014 event. *Climate Dyn.*, 52, 7339–7357. https://doi.org/10.1007/s00382-017-3531-2.

Hu, Z.-Z., Huang, B. and Pegion, K. (2008) Low cloud errors over the southeastern Atlantic in the NCEP CFS and their association with lower-tropospheric stability and air-sea interaction. *J. Geophys. Res.*, 113, doi:10.1029/2007JD009514.

Huang, B. and Hu, Z.-Z. (2007) Cloud-SST feedback in southeastern tropical Atlantic anomalous event. *J. Geophys. Res.-Oceans*, 112, C03015, doi:10.1029/2006JC003626

Hunke, E. C. (2014) Sea ice volume and age: Sensitivity to physical parameterizations and thickness resolution in the CICE sea ice model. *Ocean Model.*, 82, 45–59. doi: 10.1016/j.ocemod.2014.08.001

Hurrell, J., Meehl, G. A., Bader, D., Delworth, T. L., Kirtman, B., and Wielicki, B. (2009) A unified modeling approach to climate system prediction. *Bull. Amer. Meteor. Soc.*, 90, 1819–1832.

Hwang, Y.-T. and Frierson, D. M. W. (2013) Link between the double-Intertropical Convergence Zone problem and cloud biases over the Southern Ocean. *Proc. Natl. Acad. Sci.* USA, 110, 4935–4940, https://doi.org/10.1073/pnas.1213302110.

Im, S.-H., An, S.-I., Kim, S. T., and Jin, F.-F. (2015) Feedback processes responsible for El Niño–La Niña amplitude asymmetry. *Geophys. Res. Lett.*, 42(13), 5556–5563, https://doi.org/10.1002/2015GL064853.

Janssen, P. (2004) *The Interaction of Ocean Waves and Wind.* Cambridge: Cambridge University Press. doi:10.1017/CBO9780511525018.

Jansen, M. F., Dommenget, D. and Keenlyside, N. S. (2009) Tropical atmosphere-ocean interactions in a conceptual framework, *J. Climate*, 22(3), 550–567.

Ji, X., Neelin J. D. and Mechoso, C. R. (2015) Sea level pressure anomalies in the Western Pacific during El Niño: Why are they there? *J. Climate*, 28(22), 8860–8872.

Jiang, W., Huang, P., Li, G., and Huang, G. (2020) Emergent constraint on the frequency of central Pacific El Niño under global warming by the equatorial Pacific cold tongue bias in CMIP5/6 models. *Geophys. Res. Lett.*, 47, e2020GL089519.

Jin, E. K., et al. (2008) Current status of ENSO prediction skill in coupled ocean–atmosphere models. *Climate Dyn.*, 31, 647–64.

Jin, F.-F. (1997a) An equatorial recharge paradigm for ENSO, Part I. Conceptual model. *J. Atmos. Sci.*, 54, 811–829.

Jin, F.-F. (1997b) An equatorial recharge paradigm for ENSO, Part II: A stripped-down coupled model. *J. Atmos. Sci.*, 54, 830–847.

Jin, F.-F. and Neelin, J. D. (1993) Modes of interannual tropical ocean-atmosphere interaction — A unified view. Part I: Numerical results. *J. Atmos. Sci.*, 50, 3477–3503.

Jin, F.-F., Neelin, J. D. and Ghil, M. (1996) El Niño/Southern Oscillation and the annual cycle: Subharmonic frequency locking and aperiodicity. *Physica D*, 98, 442–465.

Jin, F. F. (1997) An equatorial ocean recharge paradigm for ENSO. Part I: Conceptual model. *J. Atmos. Sci.*, 54, 811–829.

Jin, F. F., Lin, L., Timmermann, A., and Zhao, J. (2007) Ensemble-mean dynamics of the ENSO recharge oscillator under state-dependent stochastic forcing. *Geophys. Res. Lett.*, *34*(3), 1–5.

Jin, F.-F., Chen, H.-C., Zhao, S., Hayashi, M., Karamperidou, C., Stuecker, M., and Xie, R. (2020) Simple ENSO models. In *El Niño Southern Oscillation in a Changing Climate* (eds. M. McPhaden, A. Santoso, W. Cai). American Geophysical Union, 528pp, ISBN: 978-1-119-54815-7.

Johnson, D. R. and Arakawa, A. (1996) On the scientific contributions of Professor Yale Mintz. *J. Climate*, 9, 3211–3224.

Jones, P. (1999) Conservative remapping: First- and second-order conservative remapping. *Mon. Weather Rev.*, 127, 2204–2210.

Kajtar, J. B., Santoso, A., McGregor, S., England, M. H., and Baillie, Z. (2018) Model under-representation of decadal Pacific trade wind trends and its link to tropical Atlantic bias. *Climate Dyn.*, 50, 1471–1484, doi:10.1007/s00382-017-3699-5.

Kanamitsu, M., et al. (2002) NCEP dynamical seasonal forecast system 2000. *Bull. Amer. Meteor. Soc.*, 83, 1019–1037

Kang, H.-S., Xue, Y. and Collatz, G. J. (2007) Impact assessment of satellite-derived leaf area index datasets using a general circulation model. *J. Climate*, 20(6), 993–1015.

Kang, I.-S. and Kug, J.-S. (2002) EI Niño and La Niña sea surface temperature anomalies: Asymmetry characteristics associated with their wind stress anomalies. *J. Geophys. Res. Atmospheres*, 107(19), 1–10, https://doi.org/10.1029/2001JD000393.

Kang, S. M., Held, I. M., Frierson, D. M. W., and M Zhao, M. (2008) The response of the ITCZ to extratropical thermal forcing: Idealized slab-ocean experiments with a GCM. *J. Climate*, 21(14), 3521–3532

Kasahara, A. and Washington, W. M. (1971) General circulation experiments with a six-layer NCAR model, including orography, cloudiness and surface temperature caculations. *J. Atmos. Sci.*, 28, 657–701.

Kato, S., Fred G. Rose, F. G., Rutan, D. A., Thorsen, T. J., Loeb, N. G., Doelling, D. R., Huang, X., Smith, W. L., Wenying, S. W., and Ham, S.-H. (2018) Surface irradiances of edition 4.0 clouds and the earth's radiant energy system (CERES) energy balanced and filled (EBAF) data product. *J. Climate*, 31, 4501–4527.

Keenlyside, N. S., Ding, H. and Latif, M. (2013) Potential of equatorial Atlantic variability to enhance El Niño prediction. *Geophys. Res. Lett.*, 40(10), 2278–2283.

Keenlyside, N. S., Kosaka, Y., Vigaud, N., Robertson, A. W., Wang, Y., Dommenget, D., Luo, J.-J., and Matei, D. (2021) Teleconnections in the atmosphere, In *Interacting Climates of Ocean Basins* (ed. C. R. Mechoso). Cambridge University Press.

Kerr, R. A. (2000) A North Atlantic climate pacemaker for the centuries. *Science*, 288(5473), 1984–1985.

Kessler, W. S. (2002) Is ENSO a cycle or a series of events? *Geophys. Res. Lett.*, 29, 2125, doi:10.1029/2002GL015924

Kiladis, G.cN., von Storch, H. and van Loon, H. (1989) Origin of the South Pacific Convergence Zone. *J. Climate*, 2, 1185–1195.

Kim, J.-W. and An, S.-I. (2018) Origin of early-spring central Pacific warming as the 1982–83 El Niño precursor. *Int. J. Climatol.*, 38, 2899–2906. DOI:10.1002/joc.5465.

Kim, S.-K. and An, S.-I. (2020) Untangling El Niño–La Niña asymmetries using a nonlinear coupled dynamic index. *Geophys. Res. Lett.*, 47, e2019GL085881, https://doi.org/10.1029/2019GL085881.

Kirtman, B. P., Shukla, J., Balmaseda, M., Graham, N., Penland, C., Xue, Y., and Zebiak, S. E. (2001) Current status of ENSO forecast skill: A report to the Climate Variability and Predictability (CLIVAR) Working Group on Seasonal to Interannual Prediction. WCRP Informal Report No. 23/01, 31pp.

Kleeman, R. and Moore, A. M. (1997) A theory for the limitations of ENSO predictability due to stochastic atmospheric transients. *J. Atmos. Sci.*, 54, 753–767.

Klein, S. A. and Hartmann, D. L. (1993) The seasonal cycle of low stratiform clouds. *J. Climate*, 6, 1587–1606.

Knight, J. R., Allan, R. J., Folland, C. K., Vellinga, M., and Mann, M. E. (2005) A signature of persistent natural thermohaline circulation cycles in observed climate. *Geophys. Res. Lett.*, 32(20), L20708, doi:10.1029/2005GL024233.

Konor, C. S. and Arakawa, A. (2005) Incorporation of moist processes and a PBL parameterization into the generalized vertical coordinate model. Tech. Rep. 765, Colorado State University, 75 pp. [Available online at http://hogback.atmos.col ostate.edu/pubs/PBL_tech_report_CSU_2005.pdf]

Konor, C. S., Cazes-Boezio, G., Mechoso, C. R., and Arakawa, A. (2009) Parameterization of PBL processes in an atmospheric general circulation model: Description and preliminary assessment. *Mon. Wea. Rev.*, 137(3), 1061–1082.

Koster R. D., Guo, Z., Yang, R. Dirmeyer, P. A., Mitchell, K., and Puma, M. J. (2009) On the nature of soil moisture in land surface models. *J. Climate*, 22, 4322–4335.

Koster, R. D. and Suarez, M. J. (1992) Modeling the land surface boundary in climate models as a composite of independent vegetation stands. *J. Geophys. Res.*, 97, 2697–2715.

Kug, J.-S., Jin, F.-F., Sooraj, K. P., and Kang, I.-S. (2008) State-dependent atmospheric noise associated with ENSO. *Geophys. Res. Lett.*, 35, L05701. doi:10.1029/2007GL032017

Kushnir, Y. and Held, I. M. (1996) Equilibrium atmospheric response to north Atlantic SST anomalies. *J. Climate*, 9, 1208–1220.

L'Heureux, M. L., et al. (2017) Observing and predicting the 2015/16 El Niño. *Bull. Amer. Meteor. Soc.*, 98, 1363–1382.

Larkin N. K. and Harrison, D. E. (2002) ENSO warm (El Niño) and cold (La Niña) event life cycles: Ocean surface anomaly patterns, their symmetries, asymmetries, and implications. *J. Climate*, 15, 1118–1140.

Larson, J., Jacob, R. and Ong, E. (2005) The model coupling toolkit: A new Fortran90 toolkit for building multiphysics parallel coupled models, *Int. J. High Perform. Comput. Apps.*, 19, 277–292.

Latif, M., Barnett, T. P., Cane, M. A., Flügel, M., Graham, N. E., von Storch, H., Xu, J.-S., and Zebiak, S. E. (1994) A review of ENSO prediction studies. *Climate Dyn.*, 9, 167–179.

Lee, S.-K., Mechoso, C. R., Wang, C., and Neelin, J. D. (2013) Interhemispheric influence of the northern summer monsoons on the southern subtropical anticyclones. *J. Climate*, 26, 10193–10204. doi:http://dx.doi.org/10.1175/JCLI-D-13-00106.1.

Lemarié F., Marchesiello, P. Debreu, L., and Blayo, E. (2014) Sensitivity of ocean-atmosphere coupled models to the coupling method: Example of tropical cyclone Erica. Research report RR-8651, INRIA.

Lemarié, F., Blayo, E. and Debreu, L.(2015) Analysis of ocean-atmosphere coupling algorithms: Consistency and stability, *Procedia Computer Sci.*, 51, 2066–2075, https://doi.org/10.1016/j.procs.2015.05.473, http://linkinghub.elsevier.com/retrieve/pii/S1877050915012818.

Lengaigne M., Guilyardi, E., Boulanger, J. P., Menkes, C., Delecluse, P., Inness, P., Cole, J., and Slingo, J. (2004) Triggering of El Niño by westerly wind events in a coupled general circulation model. *Climate Dyn.*, 23, 601–620.

Levine, A. F. Z. and Jin, F.-F. (2010) Noise-induced instability in the ENSO recharge oscillator. *J. Atmos. Sci.*, 67(2), 529–542. https://doi.org/10.1175/2009JAS3213.1.

Li, G. and Xie, S.-P. (2014) Tropical biases in CMIP5 multimodel ensemble: The excessive equatorial Pacific cold tongue and double ITCZ problems. *J. Climate*, 27, 1765–1780, https://doi.org/10.1175/JCLI-D-13-00337.1.

Li, H. and Misra, V. (2013) Global seasonal climate predictability in a two tiered forecast system. Part II: Boreal winter and spring seasons. *Climate Dyn.* DOI 10.1007/s00382-013-1813-x

Li, T. and Philander, S. G. H. (1997) On the annual cycle of the equatorial eastern Pacific. *J. Climate*, 9, 2986–2998.

Lindzen, R. S. and S. Nigam (1987) On the role of sea surface temperature gradients in forcing low-level winds and convergence in the tropics. *J. Atmos. Sci.*, 44, 2418–2436.

Liu, L., Zhang, C., Li, R., Wang, B., and Yang, G. (2018) C-Coupler2: A flexible and user-friendly community coupler for model coupling and nesting. *Geosci. Model Dev.*, 11, pp. 3557–3586, https://doi.org/10.5194/gmd-11-3557-2018, 2018

Lloyd, I. D. and Vecchi, G. A. (2011) Observational Evidence for Oceanic Controls on Hurricane Intensity. *J. Climate*, 24, 1138–1153

Lloyd, J., Guilyardi, E. and Weller, H. (2012) The role of atmosphere feedbacks during ENSO in the CMIP3 models. Part III: The shortwave flux feedback. *J. Climate*, 25, 4275–4293, https://doi.org/10.1175/JCLI-D-11-00178.1.

Louis, J. F. (1979) A parametric model of vertical eddy fluxes in the atmosphere. *Bound. Layer Meteor.*, 17, 187–202.

Louis, J. F., Tiedke, M. and Geleyn, J. F. (1982) A short history of the PBL parameterization at ECMWF. In *Proc. ECMWF Workshop on Boundary-Layer Parameterization*, ECMWF, Reading, UK, pp. 59–79.

Ma, C.-C., Mechoso, C. R., Robertson, A. W., and Arakawa, A. (1996) Peruvian stratus clouds and the tropical Pacific circulation: A coupled ocean–atmosphere GCM study. *J. Climate*, 9, 1635–1645, https://doi.org/10.1175/1520-0442(1996) 009,1635:PSCATT.2.0.CO;2.

Ma, J., Zhou, L. Foltz, G. R., Qu, X., Ying,, J., Tokinaga, H., Mechoso, C. R., Li, J., and Gu, X. (2020) Hydrological cycle changes under global warming and their effects on multiscale climate variability. *Ann. N. Y. Acad. Sci.*, 1472, 21–48, https://doi.org/10.1111/nyas.14335.

Madec, G. and NEMO system team (2019) NEMO ocean engine, *Scientific Notes of Climate Modelling Center* (27), Institut Pierre-Simon Laplace (IPSL).

Mahadevan, V. S., Grindeanu, I., Jacob, R., and Sarich, J. (2020) Improving climate model coupling through a complete mesh representation: A case study with E3SM (v1) and MOAB (v5.x), *Geosci. Model Dev.*, 13, 2355–2377, https://doi.org/10.5 194/gmd-13-2355-2020.

Mahajan, S., Saravanan, R. and Chang, P. (2011) The role of the wind-evaporation-sea surface temperature (WES) feedback as a thermodynamic pathway for the equatorward propagation of high-latitude sea ice–induced cold anomalies. *J. Climate*, 24, 1350–1361.

Mahrt, L. (1981) Modelling the depth of the stable boundary layer. *Bound. Layer Meteorol.*, 21, 3–19.

Maloney, E. D. and Sobel A. H. (2004) Surface fluxes and ocean coupling in the tropical intraseasonal oscillation, *J. Climate*, 17, 4368–4386.

Manabe, S. (1969) Climate and the ocean circulation: The atmospheric circulation and the hydrology of the earth's surface. *Mon. Weather Rev.*, 97, 739–774.

Manabe, S. and Bryan, K. (1969) Climate calculations with a combined ocean-atmosphere model, *J. Atmos. Sci.*, 26(4), 786–790.

Manabe, S. and Stouffer, R. J. (1980) Sensitivity of a global climate model to an increase of CO_2 concentration in the atmosphere. *J. Geophys. Res.*, 85(C10), 5529–5554.

Manabe, S. and Wetherald, R. T. (1975) The effects of doubling the CO_2 concentration on the climate of a general circulation model. *J. Atmos. Sci.*, 32, 3–15.

Manabe, S., Bryan, K. and Spelman, M. J. (1975) A global ocean-atmosphere climate model, Part I, The atmosphere circulation. *J. Phys. Oceanogr.*, 5, 3–29.

Manabe, S. and Terpstra, T. B. (1974) The effects of mountains on the general circulation of the atmosphere as identified by numerical experiments. *J. Atmos. Sci.*, 31, 3–42.

Marti, O., Nguyen, S., Braconnot, P., Valcke, S., Lemarié, F., and Blayo, E. (2021) A Schwarz iterative method to evaluate ocean-atmosphere coupling schemes: Implementation and diagnostics in IPSL-CM6-SW-VLR, *Geosci. Model Dev.* https://doi.org/10.5194/gmd-2020-307

Martín-Rey, M., Rodríguez-Fonseca, B. and Polo, I. (2015) Atlantic opportunities for ENSO prediction. *Geophys. Res. Lett.*, 42(16), 6802–6810.

Marzeion, B., Timmermann, A., Murtugudde, R., and Jin, F.-F. (2005) Biophysical feedbacks in the tropical Pacific. *J. Climate*, 18, 58–70, https://doi.org/10.1175/ JCLI3261.1.

Mason, S. J., Goddard, L., Graham, N. E., Yulaeva, E., Sun, L., and Arkin, P. A. (1999) The IRI seasonal climate prediction system and the 1997/98 El Niño event. *Bull Amer Meteor. Soc.*, 80, 1853–1873

McCreary, J. P. Jr. (1983) A model of tropical ocean-atmosphere interaction. *Mon. Wea. Rev.*, 111, 370–387.

McCreary, J. P. Jr. and Anderson, D. L. T. (1984) A simple model of El Niño and the Southern Oscillation. *Mon. Wea. Rev.*, 112, 934–946.

McGregor, S., Stuecker, M. F., Kajtar, J. B., England, M. H., and Collins, M. (2018) Model tropical Atlantic biases underpin diminished Pacific decadal variability. *Nat. Climate Change*, 8, 493–498, doi:10.1038/s41558-018-0163-4.

McPhaden, M. J. and Yu, X. (1999) Equatorial waves and the 1997-98 El Niño. *Geophys. Res. Lett.*, 26, 2961–2964.

McPhaden, M. J., Busalacchi, A. J., Cheney, R., Donguy, J.-R., Gage, K. S., Halpern, D., Ji, M., Julian, P., Meyers, G., Mitchum, G. T., Niiler, P. P., Picaut, J., Reynolds, R. W., Smith, N., and Takeuchi, K. K. (1998) The Tropical Ocean-Global Atmosphere observing system: A decade of progress. *J. Geophys. Res.*, 103(C7), 14169–14240 (1998).

Mechoso, C. R., Losada, T., Koseki, S., Mohino-Harris, E., Keenlyside, N., Rodriguez-Fonseca, B. Castaño-Tierno, A., Myers, T. A., and Toniazzo, T. (2016) Can reducing the incoming energy flux over the Southern Ocean in a CGCM improve its simulation of tropical climate? *Geophys. Res. Lett.*, 43, 11,05711,063.

Mechoso, C. R. and Arakawa, A. (2015) General Circulation Models. *Encyclopedia of Atmospheric Sciences*, 2nd Edition, Volume 4, http://dx.doi.org/10.1016/B978-0-12-382225-3.00157-2153.

Mechoso, C. R., Farrara, J. D., Drummond, L. A., Spahr, J. A., and Yu, J. Y. (2000) An atmosphere-ocean model: Code optimization and application to El Niño. *Development and Application of Computer Techniques to Environmental Studies VIII*. G. Ibarra-Berastegi, C. A. Brebbia, and P. Zannetti, Eds. WIT Press, pp. 261–278.

Mechoso, C. R., Ma, C.-C., Farrara, J. D., Spahr, J. A., and Moore, R. W. (1993) Parallelization and distribution of a coupled atmosphere-ocean general circulation model. *Mon. Wea. Rev.*, 121(7), 2062–2076.

Mechoso, C. R., Neelin, J. D. and Yu, J.-Y. (2003) Testing simple models of ENSO. *J. Atmos. Sci.*, 60, 305–318.

Mechoso, C. R., Robertson, A. W., Barth, N., Davey, M. K., Delecluse, P., Gent, P. R., Ineson, S., Kirtman, B., Latif, M., Le Treut, H., Nagai, T., Neelin, J. D., Philander, S. G. H., Polcher, J., Schopf, P. S., Stockdale, T., Suarez, M. J., Terray, L., Thual, O., and Tribbia, J. J. (1995) The seasonal cycle over the tropical Pacific in coupled ocean–atmosphere general circulation models. *Mon. Wea. Rev.*, 123, 2825–2838, https://doi.org/10.1175/1520-0493(1995)123,2825:TSCOTT.2.0.CO;2.

Medeiros, B., Williamson, D. L. and Olson J. G. (2016) Reference aquaplanet climate in the community atmosphere model, version 5. *J. Advances in Modeling Earth Sys.*, 8, 406–424, doi: 10.1002/2015MS000593

Meehl, G. A., Goddard, L., Murphy, J., Stouffer, R. J., Boer, G., Danabasoglu, G., Dixon, K., Giorgetta, M. A., Greene, A. M., Hawkins, E., Hegerl, G., Karoly, D., Keenlyside, N., Kimoto, M., Kirtman, B., Navarra, A., Pulwarty, R., Smith, D., Stammer, D., and Stockdale, T. (2009) Decadal prediction: Can it be skillful? *Bull. Amer. Meteor Soc.*, 90(10), 1467–1485.

Mellor, G. L. and Yamada, T. (1982) Development of a turbulence closure model for geophysical fluid problems. *Rev. Geophys. Space Phys.*, 20, 851–875.

Mellor, G. L. and Yamada, T. (1974) A hierarchy of turbulence closure models for planetary boundary layers. *J. Atmos. Sci.*, 31, 1791–1806 (Corrigenda, *J. Atmos. Sci.*, 34, 1482, 1977.)

Mintz, Y. (1965) Very long-term global integration of the primitive equations of atmospheric motion. In *WMO-IUGG Symposium on Research and Development Aspects of Long-Range Forecasting*, Boulder, Colo., 1964 (WMO Technical Note No. 66), edited by World Meteorological Organization [also published in *American Meteorological Society Monographs* 8 (1968): 20–36], pp. 141-55. Geneva: World Meteorological Organization.

Misra. V., Li, H., Wu, Z., and Dinapoli, S. (2014) Global seasonal climate predictability in a two tiered forecast system: Part I : Boreal summer and fall seasons. *Climate Dyn.*, 42, 1425–1448, doi:10.1007/s00382-013-1812-y

Mogensen, K. S., Magnusson, L. and Bidlot, J.-R. (2017) Tropical cyclone sensitivity to ocean coupling in the ECMWF coupled model, *J. Geophys. Res. Oceans*, 122, 4392–4412

Mohino, E., Keenlyside, N. and Pohlmann, H. (2016) Decadal prediction of Sahel rainfall: Where does the skill (or lack thereof) come from? *Climate Dyn.*, 47(11), 3593–3612.

Molteni, F. (2003) Atmospheric simulations using a GCM with simplified physical parameterizations. I: Model climatology and variability in multi-decadal experiments. *Climate Dyn.*, 20, 175–191.

NEMO TOP working group (2019) Trace in ocean paradigm (TOP) — The NEMO passive tracer engine. *Scientific Notes of Climate Modelling Center* (28), Institut Pierre-Simon Laplace (IPSL).

NEMO sea ice working group (2018) Sea ice modelling integrated initiative (SI3) — The NEMO sea ice engine. *Scientific Notes of Climate Modelling Center* (31), Institut Pierre-Simon Laplace (IPSL).

Namias, J. and Born, R. (1970) Temporal coherence in sea-surface temperature patterns. *J. Geophys. Res.*, 75, 5952–5955.

Namias, J. and Born, R. (1974) Further studies of temporal coherence in sea-surface temperature. *J. Geophys. Res.*, 79, 797–798.

Neelin, D. J. (1988) A simple model for surface stress and low-level flow in the tropical atmosphere driven by prescribed heating. *Q. J. R. Meteorol. Soc.*, 114, 747–770.

Neelin, D. J. (1990) A hybrid coupled general circulation model for El Niño studies. *J. Atmos. Sci.*, 47(5), 674–693.

Neelin, J. D. (1991) The slow sea surface temperature mode and the fast-wave limit: Analytic theory for tropical interannual oscillations and experiments in a hybrid coupled model. *J. Atmos. Sci.*, 48, 584–606.

Neelin, J. D. and Held, I. M. (1987) Modeling tropical convergence based on the moist static energy budget. *Mon. Wea. Rev.*, 115, 3–12.

Neelin, J. D., Jin, F.-F. and Syu, H.-H. (2000) Variations in ENSO phase-locking. *J. Climate*, 13, 2570–2590.

Neelin, J. D., Latif, M. and Jin, F.-F. (1994) Dynamics of coupled ocean-atmosphere models: the tropical problem. *Ann. Rev. Fluid Mech.*, 26, 617–659.

Neelin, J. D., Latif, M., Allaart, M. A. F., Cane, M. A., Cubasch, U., Gates, W. L., P. R. Gent, M. Ghil, M., Gordon, C., Lau, N. C., Mechoso, C. R., Meehl,

G. A., Oberhuber, J. M., Philander, S. G. H., Schopf, P. S., Sperber, K. R., Tokioka, T., Tribbia, J., and Zebiak, S. E. (1992) Tropical air-sea interaction in general circulation models. *Climate Dyn.*, 7, 73–104

Newman, M., Alexander, M. A., Ault, T. R., Cobb, K. M., Deser, C., Di Lorenzo, E., Mantua, N. J., Miller, A. J., Minobe, S., Nakamura, H., Schneider, N., Vimont, D. J., Phillips, A. S., Scott, J. D., and Smith, C. A. (2016) The Pacific Decadal Oscillation, revisited. *J. Climate*, 29, 4399–4427.

Nigam, S. (1997) The annual warm to cold phase transition in the eastern equatorial Pacific: Diagnosis of the role of stratus cloud-top cooling, *J. Clim.*, 10, 2447–2467.

North, G. R. (1975) Theory of energy balance climate models. *J. Atmos. Sci.*, 32, 2033–2043.

O'Neill, B. C., Kriegler, E., Ebi, K. L., Kemp-Benedict, E., Riahi, K., Rothman, D. S., van Ruijven, B. J., van Vuuren, D. P., Birkmann, J., Kok, K., Levy M. A., and Solecki, W. (2017) The roads ahead: Narratives for shared socioeconomic pathways describing world futures in the 21st century. *Global Env. Change*, 42, 169–180

Ohba, M. and Ueda, H. (2007) An impact of SST anomalies in the Indian Ocean in acceleration of the El Niño to La Niña transition. *Journal of the Meteorological Society of Japan*, 85(3), 335–348.

Okumura, Y. M. and Deser, C. (2010) Asymmetry in the duration of El Niño and La Niña. *J. Climate*, 23, 5826–5843.

Okumura, Y. M., Ohba, M., Deser, C., and Ueda, H. (2011) A proposed mechanism for the asymmetric duration of El Niño and La Niña. *J. Climate*, *24*(15), 3822–3829, https://doi.org/10.1175/2011JCLI3999.1.

Oueslati, B. and Bellon, G. (2015) The double ITCZ bias in CMIP5 models: interaction between SST, large-scale circulation and precipitation. *Climate Dyn.*, 44, 585–607, doi:10.1007/s00382-015-2468-6.

Philander, S. G. H., Gu, D., Lambert, G., Li, T., Halpern, D., Lau, N.-C., and Pacanowski, R. C. (1996) Why the ITCZ is mostly north of the equator. *J. Climate*, 9, 2958–2972

Philander, S. G. H., Hurlin, W. J. and Seigel, A. D. (1987) Simulation of the seasonal cycle of the tropical Pacific Ocean. *J. Phys. Oceanogr.*, 17, 1986–2002.

Prandtl, L. (1925) Bericht uber Untersuchungen zur ausgebildeten Turbulenz. *Ztschr. angew. Math. Mech.* (ZAMM), 5, 136–139.

Quinn, P., Beven, K. and Lamb, R. (1995) The $\ln(a/\tan\beta)$ index: How to calculate it and how to use it within the TOPMODEL framework. *Hydrol. Process.*, 9, 161–182.

Ramanathan, V., Cess, R. D., Harrison, E. F., Minnis, P., Barkstrom, B. R., Ahmad, E., and Hartmann, D. (1989) Cloud-radiative forcing and climate: Results from the Earth Radiation Budget Experiment. *Science*, 243, 57–63.

Rampal, P., Bouillon, S., Ólason, E., and Morlighem, M. (2016) NeXtSIM: A new Lagrangian sea ice model. *Cryosphere*, 10, 1055–1073. doi: 10.5194/tc-10-1055-2016.

Reichl, B. G., Hara, T. and Ginis, I. (2014) Sea state dependence of the wind stress over the ocean under hurricane winds. *J. Geophys. Res.: Oceans*, 119(1), 30–51. https://doi.org/10.1002/2013JC009289.

Reynolds, R. W., Rayner, N. A., Smith, T. M., Stokes, D. C., and Wang, W. (2002) An improved in situ and satellite SST analysis for climate. *J. Climate*, 15, 1609–1625.

Richter, I., Xie, S.-P., Wittenberg, A. T., and Masumoto, Y. (2011) Tropical Atlantic biases and their relation to surface wind stress and terrestrial precipitation. *Climate Dyn.*, 38, 985–1001, doi:10.1007/s00382-011-1038-9.

Ringler, T., Petersen, M., Higdon, R. L., Jacobsen, D., Jones, P. W., and Maltrud, M. (2013) A multi-resolution approach to global ocean modeling. *Ocean Mod.*, 69, 211–232.doi: 10.1016/j.ocemod.2013.04.010

Rodriguez-Fonseca, B., Polo, I., Garcia-Serrano, J., Losada, T., Mohino, E., Mechoso, C. R., and Kucharski, F. (2009) Are Atlantic Niños enhancing Pacific ENSO events in recent decades? *Geophys. Res. Lett.*, 36(20), L20705.

Ropelewski C. F. and Halpert, M. S. (1989) Precipitation patterns associated with the high index phase of the Southern Oscillation. *J. Climate*, 2, 268–284.

Ropelewski, C. F. and Halpert, M. S. (1986) North American precipitation and temperature associated with the El Niño Southern Oscillation (ENSO) *Mon. Wea. Rev.*, 11, 2352–2362.

Ropelewski, C. F. and Halpert, M. S. (1987) Global and regional scale precipitation patterns associated with the El Niño/Southern Oscillation. *Mon. Weather Rev.*, 115, 1606–1626.

Saha, S., Moorthi, S., Pan, H.-L., Wu, X., Wang, J., Nadiga, S., Tripp, P., Kistler, R., Woollen, J., Behringer, D., Liu, H., Stokes, D., Grumbine, R., Gayno, G., Wang, J., Hou, Y. T., Chuang, H., Juang, H.-M. H., Sela, J., Iredell, M., Treadon, R., Kleist, D., Van Delst, P., Keyser, D., Derber, J., Ek, M., Meng, J., Wei, H., Yang, R., Lord, S., van den Dool, H., Kumar, A., Wang, W., Long, C., Chelliah, M., Xue, Y., Huang, B., Schemm, J. -K., Ebisuzaki, W., Lin, R., Xie, P., Chen, M., Zhou, S., Higgins, W., Zou, C.-Z., Liu, Q., Chen, Y., Han, Y., Cucurull, L., Reynolds, R. W., Rutledge, G., and Goldberg, M. (2010) The NCEP climate forecast system reanalysis. *Bull. Amer. Meteor. Soc.*, 91, 1015–1058.

Sasaki, W., Doi, T., Richards, K. J., and Masumoto, Y. (2014) Impact of the equatorial Atlantic sea surface temperature on the tropical Pacific in a CGCM. *Climate Dyn.*, 43, 2539–2552, doi:10.1007/s00382-014-2072-1.

Schneider, S. H. and Harvey, L. D. D. (1985) Computational efficiency and accuracy of methods for asynchronously coupling atmosphere-ocean climate models. Part I: Testing with mean annual model. *J. Phys. Oceanogr.*, 16, 3–10.

Schopf, P. S. and Suarez, M. J. (1988) Vacillations in a coupled ocean-atmosphere model. *J. Atmos. Sci.*, 45, 549–566.

Schubert, S., Stewart, R., Wang, H., Barlow, M., Berbery, H., Cai, W., Hoerling, M., Kanikicharla, K., Koster, R., Lyon, B., Mariotti, A., Mechoso, C. R., Müller, O., Rodriguez-Fonseca, B., Seager, R., Seneviratne, S. I., Zhang, L., and Zhou, T. (2016) Global meteorological drought: A synthesis of current understanding with a focus on SST drivers of precipitation. *J. Climate*, 29, 3989–4019, DOI:10.1175/JCLI-D-15-0452.1

Seager, R., Cane, M., Henderson, H., Lee, D. E., Abernathey, R., and Zhang H. (2019) Strengthening tropical Pacific zonal sea surface temperature gradient consistent with rising greenhouse gases. *Nature Climate Change*, 9, 517–522.

Seibert, P., Beyrich, F., Gryning, S. E., Joffre, S., Rasmussen, A., and Tercier, P. (2000) Review and intercomparison of operational methods for the determination of the mixing height, *Atmos. Environ.*, 34, 1001–1027.

Seidel, D. J., Zhang, Y., Beljaars, A., Golaz, J.-C., Jacobson, A. R., and Medeiros, B. (2012) Climatology of the planetary boundary layer over the continental United States and Europe, *J. Geophys. Res.*, 117, D17106, doi:10.1029/2012JD018143.

Sellers, W. D. (1969) A global climatic model based on the energy balance of the earth-atmosphere system. *J. Appl. Meteor.*, 8, pp. 392–400.

Sellers, P. J., Randall, D. A., Collatz, G. J., Berry, J. A., Field, C. B., Dazlich, D. A., Zhang, C., Collelo, G. D., and Bounoua, L. (1996) A revised land surface parameterization (SiB2) for atmospheric GCMs, Part 1: Model formulation, *J. Climate*, 9, pp. 676–705.

Seo, J., Kang, S. M. and Frierson, D. M. W. (2014) Sensitivity of intertropical convergence zone movement to the latitudinal position of thermal forcing. *J. Climate*, 27, 3035–3042.

Seth, A., Giorgi, F. and Dickinson, R. E. (1994) Simulating fluxes from heterogeneous land surfaces: Explicit subgrid method employing the Biosphere–Atmosphere Transfer Scheme (BATS). *J. Geophys. Res.*, 99, 18651–18668.

Sheen, K. L., Smith, D. M., Dunstone, N. J., Eade, R., Rowell, D. P., and Vellinga, M. (2017) Skilful prediction of Sahel summer rainfall on inter-annual and multi-year timescales. *Nature Comm.*, 8, 14966.

Shukla, J. and Mintz, Y. (1982) Influence of land-surface evapotranspiration on the earth's climate. *Science*, 215, 1498–1501.

Sklower, K., Robinson, H., Mechoso, C. R., Drummond, L. A., Spahr, J. A., and Farrara, J. D. (2002) The Distributed Data Broker: A decentralized mechanism for periodic exchange of fields between multiple ensembles of parallel computations. Technical Report, Computer Science Dept, University of California, Berkeley, 45 pp.

Smith, G. C., Bélanger, J.-M., Roy, F., Pellerin, P., Ritchie, H., Onu, K., Roch, M., Zadra, A., Colan, D. S., Winter, B., Fontecilla, J.-S., and Deacu, D. (2018) Impact of coupling with an ice–ocean model on global medium-range NWP forecast skill. *Mon. Wea. Rev.*, DOI: 10.1175/MWR-D-17-0157.1

Smith, R., Dukowicz, J. and Malone, R. (1992) Parallel ocean general circulation modeling. *Physica D*, 60, 38–61.

Smith, R., Jones, P., Briegleb, B., Bryan, F., Danabasoglu, G., Dennis, H., Dukowicz, J., Eden, C., Fox-Kemper, B., Gent, P. R., Hecht, M., Jayne, S., Jochum, M., Large, W. G., Lindsay, K., Maltrud, M., Norton, N. J., Peacock, S. L., Vertenstein, M., and Yeager, S. (2010) *The Parallel Ocean Program (POP) Reference Manual: Ocean Component of the Community Climate System Model (ccsm) and Community Earth System Model (CESM).* Los Alamos Technical Report No. LAUR-10-01853, 140. Los Alamos National Laboratory.

Song, X. and Zhang, G. (2018) The roles of convection parameterization in the formation of double ITCZ syndrome in the NCAR CESM, Part I: Atmospheric Processes. *J. Adv. Model. Syst.* doi.org/10.1002/2017MS001191.

Stevens, B. (2002) Entrainment in stratocumulus-topped mixed layers. *Q. Royal Meteor. Soc.*, 128, 2663–2690.

Stieglitz, M., Rind, D., Famiglietti, J., and Rosenzweig, C. (1997) An efficient approach to modeling the topographic control of surface hydrology for regional and global climate modeling. *J. Climate*, 10, 118–137.

Straume, A. G., N'Dri Koffi, E. and Nodop, K. (1998) Dispersion modelling using ensemble forecasts compared to ETEX measurements, *J. Appl. Meteorol.*, 37, 1444–1155.

Stockdale, T., Alonso-Balmaseda, M., Johnson, S., Ferranti, L., Molteni, F., Magnusson, L., Tietsche, S., Vitart, F., Decremer, D., Weisheimer, A., Roberts, C. D., Balsamo, G. Keeley, S., Mogensen, K., Zuo, H., Mayer, M., and Monge-Sanz,

B. M. (2018) SEAS5 and the future evolution of the long-range forecast system. European Centre for Medium Range Weather Forecasts (ECMWF), Technical Memorandum 835, 81 pp.

Su, J., Zhang, R., Li, T., Rong, X., Kug, J. S., and Hong, C. C. (2010) Causes of the El Niño and La Niña amplitude asymmetry in the equatorial Eastern Pacific. *J. Climate*, 23(3), pp. 605–617. https://doi.org/10.1175/2009JCLI2894.1.

Suarez, M. J. and Schopf, P. S. (1988) A delayed action oscillator for ENSO. *J. Atmos. Sci.*, 45, 3283–3287.

Suarez, M. J., Arakawa, A. and Randall, D. A. (1983) The parameterization of the planetary boundary layer in the UCLA general circulation model: Formulation and results. *Mon. Wea. Rev.*, 111, 2224–2243.

Sud, Y. C., Shukla, J. and Mintz, Y. (1988) Influence of land surface roughness on atmospheric circulation and precipitation: A sensitivity study with a general circulation model. *J. Appl. Meteor.*, 27, pp. 1036–1054.

Syu, H.-H. and Neelin, J. D. (2000) ENSO in a hybrid coupled model. Part II: Prediction with piggyback data assimilation. *Climate Dyn.*, 16, 35–48

Syu, H.-H., Neelin, J. D. and Gutzler, D. (1995) Seasonal and interannual variability in a hybrid coupled GCM. *J. Climate*, 8, 2121–2143.

Tang, Y. and Hsien, W. W. (2002) Hybrid coupled models of the tropical Pacific — II ENSO prediction. *Climate Dyn.*, 19, 343–353.

Taschetto, A. S., Sen Gupta, A., Jourdain, N. C., Santoso, A., Ummenhofer, C. C., and England, M. H. (2014) Cold tongue and warm pool ENSO events in CMIP5: Mean state and future projections. *J. Climate*, 27, 2861–2885.

Tautges, T. J., Meyers, R., Merkley, K., Stimpson, C., and Ernst, C. (2004) MOAB: A Mesh-Oriented datABase, SAND2004-1592, Sandia National Laboratories.

Taylor, P. K. and Yelland, M. J. (2001) The dependence of sea surface roughness on the height and steepness of the waves. *J. Phys. Oceanogr.*, 31, 572–590.

Terray, L., Thual, O., Belamari, S., Déqué, M., Dandin, P., Lévy, C., and Delecluse, P. (1995) Climatology and interannual variability simulated by the ARPEGE-OPA model, *Climate Dyn.*, 11, 487–505.

Theurich, G., Deluca, C., Campbell, T., Liu, F., Saint, K., Vertenstein, M., Chen, J., Oehmke, R., Doyle, J., Whitcomb, T., Wallcraft, A., Iredell, M., Black, T., Da Silva, A. M., Clune, T., Ferraro, R., Li, P., Kelley, M., Aleinov, I., Balaji, V., Zadeh, N., Jacob, R., Kirtman, B., Giraldo, F., McCarren, D., Sandgathe, S., Peckham, S., and Dunlap IV, R. (2016) The earth system prediction suite: Toward a coordinated U.S. modeling capability, *Bull. Amer. Meteor. Soc.*, 97, 1229–1247, https://doi.org/10.1175/BAMS-D-14-00164.1.

Timmermann, A. and Jin, F.-F. (2002) Phytoplankton influences on tropical climate. *Geophys. Res. Lett.*, 29, 19-1-19-4, https://doi.org/10.1029/2002GL015434.

Timmermann, A., An, S.-I., Kug, J.-S., Jin, F.-F., Cai, W., Capotondi, A., Cobb, K. M., Lengaigne, M., McPhaden, M. J., Stuecker, M. F., Stein, K., Wittenberg, A. T., Yun, K.-S., Bayr, T., Chen, H.-C., Chikamoto, Y., Dewitte, B., Dommenget, D., Grothe, P., Guilyardi, E., Ham, Y.-G., Hayashi, M., Ineson, S., Kang, D., Kim, S., Kim, W. M., Lee, J.-Y., Li, T., Luo, J.-J., McGregor, S., Planton, Y., Power, S., Rashid, H., Ren, H.-L., Santoso, A., Takahashi, K., Todd, A., Wang, G., Wang, G., Xie, R., Yang, W.-H., Yeh, S.-W., Yoon, J., Zeller, E., and Zhang, X. (2018) El Niño-Southern Oscillation complexity, *Nature*, 559, 535–545.

Ting, M., Kushnir, Y., Seager, R., and Li, C. (2009) Forced and internal twentieth-century SST trends in the North Atlantic. *J. Climate*, 22, 1469–1481.

Tokinaga, H., Xie, S.-P., Deser, C., Kosaka, Y., and Okumura, Y. M. (2012) Slowdown of the Walker circulation driven by tropical Indo-Pacific warming. *Nature*, 491, 439–443.

Toniazzo, T. and Woolnough, S. (2014) Development of warm SST errors in the southern tropical Atlantic in CMIP5 decadal hindcasts. *Climate Dyn.*, 43, 2889–2913, doi:10.1007/s00382-013-1691-2

Troen, I. and Mahrt, L. (1986) A simple model of the atmospheric boundary layer: Sensitivity to surface evaporation. *Bound. Layer Meteor.*, 37, 129–148.

Trzaska, S., Robertson, A. W., Farrara, J. D., and Mechoso C. R. (2007) South Atlantic variability arising from air-sea coupling: local mechanisms and tropical–subtropical interactions. *J. Climate*, 20, 3345–3365.

Ullrich, P. A. and Taylor, M. A. (2015) Arbitrary-order conservative and consistent remapping and a theory of linear maps: Part I, *Mon. Wea. Rev.*, 143, 2419–2440.

Ullrich, P. A., Devendran, D. and Johansen, H. (2016) Arbitrary-order conservative and consistent remapping and a theory of linear maps: Part II, *Mon. Weather Rev.*, 144, 1529–1549.

Valcke, S. (2013) The OASIS3 coupler: A European climate modelling community software, *Geosci. Model Dev.*, 6, 373–388, https://doi.org/10.5194/gmd-6-373-2013.

Valcke, S., Balaji, V., Craig, A., DeLuca, C., Dunlap, R., Ford, R. W., Jacob, R., Larson, J., O'Kuinghttons, R., Riley, G. D., and Vertenstein, M. (2012) Coupling technologies for Earth System Modelling, *Geosci. Model Dev.*, 5, 1589–1596, https://doi.org/10.5194/gmd-5-1589-2012.

Valcke, S. and Guilyardi, E. (2008) On a revised ocean-atmosphere physical coupling interface and about technical coupling software. ECMWF Workshop on Ocean-Atmosphere Interactions, 10-12 November 2008

Vecchi, G. A. and Soden, B. J. (2007) Global warming and the weakening of the tropical circulation. *J. Climate*, 20, 4316–4340.

Verbickas, S. (1998) Westerly wind bursts in the tropical Pacific. *Weather*, 53, 282–284.

Vimont, D. J., Battisti, D. S. and Hirst, A. C. (2001) Footprinting: A seasonal connection between the tropics and mid-latitudes. *Geophys. Res. Lett.*, 28, 3923–3926.

Vimont, D. J., Wallace, J. M. and Battisti, D. S. (2003) The seasonal footprinting mechanism in the Pacific: Implications for ENSO. *J. Climate*, 16, 2668–2675.

Vintzileos, A. and Sadourny, R. (1997) A general interface between an atmospheric general circulation model and underlying ocean and land surface models: Delocalized physics scheme. *Mon. Wea. Rev.*, 125, 926–941.

Voldoire, A., Claudon, M., Caniaux, G., Giordani, H., and Roehrig, R. (2014) Are atmospheric biases responsible for the tropical Atlantic SST biases in the CNRM-CM5 coupled model?. *Climate Dyn.*, 43(11), 2963–2984.

Walker, G. T. (1923) Correlation in seasonal variations of weather VIII. Mem. India Meteor. Dept., 24, pp. 75–131.

Wallace, J. M., Rasmusson, E. M., Mitchell, T. P., Kousky, V. E., Sarachik, E. S., and von Storch, H. (1998) On the structure and evolution of ENSO-related climate variability in the tropical Pacific: Lessons from TOGA. *J. Geophys. Res.*, 103(C7), 14,241–14,259.

Wan, H., Giorgetta, M. A., Zängl, G., Restelli, M., Majewski, D., Bonaventura, L., Fröhlich, K., Reinert, D., Rípodas, P., Kornblueh, L., and Förstner, J. (2013) The ICON-1.2 hydrostatic atmospheric dynamical core on triangular grids – Part 1: Formulation and performance of the baseline version, *Geosci. Model Dev.*, 6, 735–763, https://doi.org/10.5194/gmd-6-735-2013.

Wang, B., Luo, X., Yang, Y.-M., Sun, W., Cane, M. A., Cai, W., Yeh, S.-W., and Liu, J. (2019) Historical change of El Niño properties sheds light on future changes of extreme El Niño. *Proc. Natl. Acad. Sci.*, 116, 22512–22517

Wang, B., Wu, R. and Lukas, R. (1998) Roles of the western north Pacific wind variation in thermocline adjustment and ENSO phase transition. *J. Met. Soc. Japan*, 77, 1–16.

Wang, C. (2001a) A unified oscillator model for the El Niño-Southern Oscillation. *J. Climate*, 14, 98–115.

Wang, C. (2001b) On the ENSO mechanisms. *Advances in Atmospheric Sciences*, 18, 674–691.

Wang, C., Zhang, L., Lee, S. K., Wu, L., and Mechoso, C. R. (2014) A global perspective on CMIP5 climate model biases. *Nat. Climate Change*, 4, 201–205, https://doi.org/10.1038/nclimate2118.

Wang, Z., Schneider, E. K. and Burls N. J. (2019) The sensitivity of climatological SST to slab ocean model thickness. *Climate Dyn.*, 53, 5709–5723.

Washington, W. M., Weatherly, J. W. Meehl, G. A. Semtner Jr., A. J. Bettge, T. W. Craig, A. P. Strand Jr., W. G. Arblaster, J. Wayland, V. B. James, R., and Zhang, Y. (2000) Parallel climate model (PCM) control and transient simulations. Climate Dynamics, 16, pp. 755–774.

Washington, W. M., Semtner Jr., A. J., Meehl, G. A., Knight, D. J., and Mayer, T. A. (1980) A general circulation experiment with a coupled atmosphere, ocean and sea ice model, *J. Phys. Oceanogr.*, 10, 1887–1908, http://dx.doi.org/10.1175/152 0-0485(1980)010⟨1887:AGCEWA⟩2.0.CO;2.

Washington, W. M. and Meehl, G. A. (1984) Seasonal cycle experiment on the climate sensitivity due to a doubling of CO2 with an atmospheric general circulation model coupled to a simple mixed-layer ocean model. *J. Geophys. Res.*, 89, 9475–9503.

Watanabe, M., Kug, J.-S., Jin, F.-F., Collins, M., Ohba. M., and Wittenberg, A. T. (2012) Uncertainty in the ENSO amplitude change from the past to the future. *Geophys. Res. Lett.*, 39, L20703

Weisberg, R. H., Wang, C. and Virmani, J. I. (1999) Western Pacific interannual variability associated with the El Niño–Southern Oscillation. *J. Geophys. Res.*, 104, 5131–5149.

Weisberg, R. H. and Wang, C. (1997) A western Pacific oscillator paradigm for the El Niño-Southern Oscillation. *Geophys. Res. Lett.*, 24, 779–782.

Wetherald, R. T. and Manabe, S. (1988) Cloud feedback processes in a general circulation model. *J. Atmos. Sci.*, 45(8), 1397–1415.

Wetzel, P. J. (1982) Toward parameterization of the stable boundary layer. *J. Appl. Meteor.*, 21, pp. 7–13.

Wyrtki, K. (1975) El Niño—the dynamic response of the equatorial Pacific Ocean to atmospheric forcing. *J. Phys. Oceanogr.*, 5, 572–584.

Xiang, B., Zhao, M., Held, I. M., and Golaz, J. C. (2017) Predicting the severity of spurious "double ITCZ" problem in CMIP5 coupled models from AMIP simulations. *Geophys. Res. Lett.*, 44, 1520–1527, https://doi.org/10.1002/2016G L071992.

Xie, P. and Arkin P. A. (1997) Global precipitation: A 17-year monthly analysis based on gauge observations, satellite estimates, and numerical model outputs. *Bull. Amer. Meteor. Soc.*, 78, 2539–2558.

Xie, S.-P. (2020) Ocean warming pattern effect on global and regional climate change. *AGU Advances*, 1, e2019AV000130.

Xie, S.-P., Deser, C., Vecchi, G. A., Ma, J., Teng, H., and Wittenberg, A. (2010) Global warming pattern formation: Sea surface temperature and rainfall. *J. Climate*, 23, 966–986.

Xu, K., Tam, C.-Y., Zhu, C., Liu, B., and Wang, W. (2017) CMIP5 projections of two types of El Niño and their related tropical precipitation in the twenty-first century. *J. Climate*, 30, 849–864. https://doi.org/10.1175/jcli-d-16-0413.1

Xue, Y., De Sales, F., Vasic, R., Mechoso C. R., Arakawa A., and Prince, S. (2010) Global and seasonal assessment of interactions between climate and vegetation biophysical processes: A GCM study with different land–vegetation representations. *J. Climate*, 23, 1411–1433.

Xue, Y., Sellers, P. J., Kinter III, J. L., and J. Shukla, J. (1991) A simplified biosphere model for global climate studies. *J. Climate*, 4, 345–364.

Yang, Y.-M., An, S.-I., Wang, B., and Park, J. H. (2020) A global-scale multidecadal variability driven by Atlantic multidecadal oscillation. *National Science Review*, 7, 1190–1197.

Yasunari, T., Saito, K. and Takata, K. (2006) Relative roles of large-scale orography and land surface processes on global hydro-climate. Part I. Impacts on monsoon systems and tropics. *J. Hydrometeor.*, 7, 626–641.

Yeager, S. G., Danabasoglu, G., Rosenbloom, N. A., Strand, W., Bates, S. C., Meehl, G. A., Karspeck, A. R., Lindsay, K., Long, M. C., Teng, H., and Lovenduski, N. S. (2018) Predicting near-term changes in the earth system: A large ensemble of initialized decadal prediction simulations using the Community Earth System Model. *Bull. Amer. Meteor. Soc.*, 99(9), 1867–1886.

Yeh, S. W., Kug, J. S., Dewitte, B., Kwon, M. H., Kirtman, B. P., and Jin, F.-F. (2009) El Niño in a changing climate. *Nature*, 461, 511–514.

Yu, J.-Y., Campos, E., Du, Y., Eldevik, T., Gille, S. T., Losada, T., McPhaden, M. J., and Smedsrud, L. H. (2021) Variability of the Oceans. In *Interacting Climates of Ocean Basins* (ed. C. R. Mechoso). Cambridge University Press, Cambridge, UK and New York, NY, USA.

Yu, J.-Y. and Mechoso, C. R. (2001) A coupled atmosphereocean GCM study of the ENSO cycle. *J. Climate*, 14, 2329–2350.

Yu, L. A. and Rienecker, M. M. (1998) Evidence of an extratropical atmospheric influence during the onset of the 1997–98 El Niño. *Geophys. Res. Lett.*, 25, 3537–3540.

Yu, L., Jin, X. and Weller, R. A. (2008) Multidecade Global Flux Datasets from the Objectively Analyzed Air-sea Fluxes (OAFlux) Project: Latent and sensible heat fluxes, ocean evaporation, and related surface meteorological variables. Woods Hole Oceanographic Institution, OAFlux Project Technical Report. OA-2008-01, 64pp.

Yu, L., Weller, R. A. and Liu, T. W. (2003) Case analysis of a role of ENSO in regulating the generation of westerly wind bursts in the western equatorial Pacific. *J. Geophys. Res.*, 108, 3128, doi:10.1029/2002JC001498.

Zebiak, S. (1993) Air-sea interaction in the equatorial Atlantic region. *J. Climate*, 6, 1567–1586.

Zebiak, S. E. (1984) Thesis (Ph.D.), Massachusetts Institute of Technology, Dept. of Earth, Atmospheric, and Planetary Sciences, 1985.

Zebiak, S. E. (1986) Atmospheric convergence feedback in a simple model for El Niño. *Mon. Wea. Rev.*, 114, 1263–1271.

Zebiak, S. E. and Cane, M. A. (1987) A model El Niño-Southern Oscillation. *Mon. Weather Rev.*, 115, 2262–2278.

Zermeño-Diaz, D. M. and Zhang, C. (2013) Possible root causes of surface westerly biases over the equatorial Atlantic in global climate models. *J. Climate*, 26, 8154–8168, doi:10.1175/JCLI-D-12-00226.1.

Zhang, R. and Delworth, T. L. (2006) Impact of Atlantic multidecadal oscillations on India/Sahel rainfall and Atlantic hurricanes. *Geophys. Res. Lett.*, 33, L17712, doi:10.1029/2006GL026267.

Zhang, R.-H. (2015) A hybrid coupled model for the Pacific ocean–atmosphere system. Part I: Description and basic performance. *Adv. Atmos. Sci.*, 32(3), 301–318. doi: 10.1007/s00376-014-3266-5.

Zhang, R.-H. and Busalacchi, A. J. (2009a) Freshwater flux (FWF)-induced oceanic feedback in a hybrid coupled model of the tropical Pacific. *J. Climate*, 22, 853–879.

Zhang, R.-H. and Busalacchi, A. J. (2009b) An empirical model for surface wind stress response to SST forcing induced by Tropical Instability Waves (TIWs) in the eastern equatorial Pacific. *Mon. Wea. Rev.*, 137, 2021–2046.

Zhang, Y., Wallace, J. M. and Battisti, D. S (1997) ENSO-like interdecadal variability: 1900-93. *J. Climate*, 10, 1004–1020.

Zhao, X. and Li, J. (2010) Winter-to-winter recurrence of SSTA in the Northern Hemisphere, *J. Climate*, 23, 3835–3854.

Zilitinkevich, S. S. (1972) On the determination of the height of the Ekman boundary layer. *Bound. Layer Meteorol.*, 3, 141–145

Zilitinkevich, S., Baklanov, A., Rost, J, Smedman, A.-S., Lykosov, V., and Calanca, P. (2002) Diagnostic and prognostic equations for the depth of the stably stratified Ekman boundary layer. *Q. J. R. Meteor. Soc.*, 128, 25–46.

Zuidema, P., Chang, P., Medeiros, B., Kirtman, B., Mechoso, C. R., Schneider, E., Toniazzo, T., Richter, I., Small, J., Bellomo, K., Brandt, P., de Szoeke, S., Farrar, T., Jung, E., Kato, S., Li, M., Patricola, C., Wang, Z., Wood, R., and Xu, Z. (2016) Challenges and prospects for reducing coupled climate model SST biases in the eastern tropical Atlantic and Pacific Oceans: The U.S. CLIVAR Eastern Tropical Oceans Synthesis Working Group. *Bull. Amer. Meteor. Soc.*, 97, 2305–2328, https://doi.org/10.1175/BAMS-D-15-00274.1.

Index

www.ingramcontent.com/pod-product-compliance
Lightning Source LLC
Chambersburg PA
CBHW081105220326
41598CB00038B/7233